1974

book may be kept

The Mind of Man

*some views
and a theory of
cognitive development*

The Mind of Man

some views
and a theory of
cognitive development

G. Thomas Rowland

NEW YORK UNIVERSITY AND
THE INSTITUTE FOR EPISTEMIC STUDIES

J. Carson McGuire

Prentice-Hall, Inc., Englewood Cliffs, N.J.

Cover photograph by Richard Traub

Copyright © 1971 by Prentice-Hall, Inc.
Englewood Cliffs, New Jersey

Prentice-Hall International, Inc., *London*
Prentice-Hall of Australia Pty. Ltd., *Sydney*
Prentice-Hall of Canada Ltd., *Toronto*
Prentice-Hall of India Private Limited, *New Delhi*
Prentice-Hall of Japan, Inc., *Tokyo*

Dedicated to

the brilliant men
whose works are discussed
in this small volume,

the great behavioral scientists
who took the first giant steps, and
sometimes suffered for their efforts,

and the students,
who now and in the future
bear the awesome responsibility of
continuing within the framework of science
the task set before them.

In Memoriam

Professor J. Carson McGuire
1910–1969

Fellow of the
American Psychological Association

Co-founder of the
Institute for Epistemic Studies

Senior Professor
Department of Educational Psychology
The University of Texas at Austin

He taught.

Contents

Foreword

It is a pleasure and an honor to be invited to write a foreword to this timely and important book. Timely, because it comes at the brink of a new decade, a time of governmental reappraisal of its commitment to the schools, a time of crisis for the American system of public education, a time for courageous, skillful educational leadership as these tests are encountered. Important, for during no previous era has the need for unity and sharing between the behavioral sciences (including education) been so urgent or so formidable a challenge.

Despite the warning signals currently emanating from the behavioral sciences, the tone of this book is definitely optimistic. The choice of content, the selection of minds, the implications for educational practice, all of these reflect the faith of the writers in the intelligence of man and his ability to bring order to disorder through the methods of science. There is no dwelling herein upon what behavioral scientists *cannot do* nor upon what children *cannot become*. Such a direction would be inconsistent with the great minds whose works are represented here. And there is no implication that science is cold and contrary to the essential humanness of man. Quite the contrary, for the sensitivity and skill of science may well hold the key to the needed changes in educational process, particularly in the schools of our nation.

The writers exhibit rare analytical skill in their dissection of the voluminous works represented here and the result is a system, not a smorgasbord. The emerging set of action oriented principles can be traced through the preceding works of eminent psychologists, to the defining of a theoretical model for the *science* of *developmental education,* a behavioral science utilizing the principles of psychology and education.

Remarkable scientific insights and experimentations clarified *invariant sequence* in the development of intelligent behavior, and stimulated the notion that experience, though insufficient in itself, is the central mediating factor in intellectual development. The logically consistent view that *time per se* is of no educational importance, except in the sense that *learning takes time,* places a rather serious indictment against such current practices in public schools as grade level standards, mandatory entrance ages, and "reading readiness." This indictment is further strengthened by the scientific principle that the rate and timing of intellectual development is *highly variant,* unique to the individual; not an original idea, but an idea confined largely to educational theory and foreign to present instructional practice in most American schools.

Several critical dimensions of a theory of instruction, structure, process, motivation, and reinforcement are clarified in varying degrees through a synthesis of represented works. Evidence for a cognitive hierarchy (*structure*) is a recurring thread among the works of Piaget, White, Bruner, and Hunt. The acquisition of knowledge is an ordinal (cumulative) *process* reciprocated by organism-environment interactions, not predetermined, and not a linear product of maturation. The author's description of a *structure-process approach* or *system* is an extremely useful section for educators, drawing together as it does the complex psychological bases into a logical framework for curriculum development. Berlyne, White, and Hunt provide the bases for many of the needed applications of *motivation* into classrooms. One promising strategy, discovery learning, is explored, but a great deal of future exploration will be required to build sound schemes into instructional practice.

Perhaps the most important feature of this book is the consistent emphasis upon an integration of the behavioral sciences, psychology and education, emerging as developmental education. For psychology can "envision the possible" and education can formulate hypotheses and put them to the test. When such interdisciplinary skills are applied extensively to practice, the practitioner will be able to clearly express the goals of instruction and individualized teaching may for the first time become fact instead of fantasy.

Finally I express my compliments to Professor Rowland for this important contribution to psychology and education; and I join the many friends, colleagues, and ex-students of the late Professor McGuire in extending my respect to this great humanitarian.

JOE L. FROST
*The University of Texas at Austin
and The Institute for Epistemic Studies*

January, 1970

Preface

The chapters of this book have appeared in a similar form over a considerable period of time in three journals, *Psychology in the Schools*, *Journal of School Psychology*, and *Educational Leadership*. These articles were co-authored with Prof. Carson McGuire, recently deceased, who helped make the editorial changes for this volume. In most cases the changes were for the sake of clarity; in addition, the first and last chapters were added because the authors believed that the inclusion of these articles would contribute to the understanding of human intelligent behavior from the context of the *Structure-Process* system.

It may seem that the internal consistency of much of the content of cognitive psychology could be called "obvious" by a philosopher. Fortunately, there is no such category in psychology, since as a science, psychology depends largely upon the scientist's conceptualizations as they are supported by sometimes tedious and always time-consuming research. The same, of course, should be true of education, a field in which a gradual awareness of the dangers of "artistic" teaching seems to be developing.

Much of the research which supports the ideas presented in this book is cited only briefly, but it is expected that the interested student will use the references listed in the bibliography as a starting point to satisfy his individual needs. In this manner, the student of cognition can confirm for himself the accuracy of some of the theories and concepts discussed here. These theoretical concepts, models, and ideas have been presented for the benefit of students of cognitive development within the sciences of psychology and education, as well as for colleagues involved in the field of cognitive psychology, which has been sometimes called genetic epistemology.

Genetic Epistemology

Genetic epistemology is a somewhat recent arrival on the scene of American education and psychology, having stirred only a brief flurry of interest in the early twentieth century. The belated renewal of interest in this area is perhaps partly because some of the earlier speculations of epistemologists were too introspective and philosophical, tending sometimes to be almost theological in nature. Such considerations found little sympathy or understanding among American scientists; therefore, it is not surprising that the renewal of psychological interests in genetic epistemology sought a more favorable climate in Europe.

In Europe the great introspectionist Wundt was able to differentiate the science of psychology once and for all from philosophy; in Europe, Freud found a favorable intellectual climate for his insights; and, in Europe the wide-ranging area known as *gestalt* psychology was born. In America, Lewin's Field Theory began to express topologically *gestalt* notions, but his death left this expression incomplete, until the advent of J. P. Guilford.

The American version of genetic epistemology is behaviorist in the truest sense of the definition—as is shown in Chapter 3 of this book on the work of Daniel E. Berlyne—but perhaps this new or recent development should be more correctly identified with that peculiarly American school of psychology known as *functionalism,* early represented in the purposive behaviorism of Tolman, whose work seems to have profoundly influenced Jerome S. Bruner.

The Genevan school of genetic epistemology, represented by the work of Jean Piaget and some of his colleagues, such as Bärbel Inhelder in Geneva and Gilbert Voyat in the United States, seems little interested in the scientific aspects of American behaviorism, and little concerned with the problems of education. In spite of Berlyne's brilliant translation of some of Piaget's thoughts into behavioral terms (a work which can be compared with the efforts of John Dollard and Neal Miller in relation to the concepts of Freud), Piaget has shown only passing interest in the behavioral sciences of psychology and education. This lack of interest shown by the acknowledged leader of the Genevan school seems unfortunate, especially since his unique and often individualized methodologies have made quantitative analyses of an otherwise brilliant approach to intelligent behavior exceptionally difficult.[1] The conservation studies which have typified Piaget's work and that of his associates have been elaborately conceptualized as indices of the development of intelligent behavior, and have been widely replicated, especially in the United States.

The disinterest, or possibly distrust, of the Genevans, however, may well be the fault of American psychologists such as Bandura and McDonald (1963) who attempted to infirm the concept of sequential, individualized development by utilizing what appears to be an inappropriate research model, demographic methods of quantitative analysis, and possibly by extrapolating somewhat beyond the data. The Bandura (Stanford University) and McDonald (New York University) study has been discussed fully and brought into serious question by other authorities (Kelly and Cody, 1969, pp. 234–39) who appear to confirm our belief that this type of comparative analysis has not yielded constructive interdisciplinary communication. In an earlier text, McDonald (1959) emphasized the importance of peer interaction and group decision making. For example, he says, "When individuals have an opportunity to participate in the decision-making process, . . . behavior changes are more frequent" (p. 537). In his chapter on "The School as a Social System" (pp. 548–76), he again stresses the influences of peers, parents or authority figures, and teachers in the individual's development of intelligent behavior. This is a developmental concept which seems quite compatible with the model proposed in the final chapter of this book, which we have called the

[1] Some important breakthroughs in this area seem very near, especially following the Conference on Ordinal Scales of Cognitive Development, sponsored by California Test Bureau, a division of McGraw-Hill Book Company, from February 9–11, 1969 in Carmel, California. Some leaders in these efforts are Prof. Beth Stephens (Temple University, Philadelphia), Prof. Gene Glass (University of Colorado), Prof. Remy Droz (University of Geneva), and Prof. Peter Bentler (University of California, Los Angeles).

dyadic model. The dyadic model proposes that the development of intelligent behavior transpires in a cultural context (Frost and Rowland, 1969, p. 12; Rowland and McGuire, 1969). This model respects operant conditioning in certain instances, but conceives of such strategies as preliminary presentations of relevant data to learners rather than the exercise of decision-making power by some authority figure such as the teacher or psychologist.

The conservation replications, and some of the innovative additions, are unquestionably meritorious; nevertheless, they are hardly sufficient evidence to explain the origins, development, and education of the human mind. Perhaps, it is too much to expect that a brilliant theorist such as Piaget, who has concentrated on the origins of the operations of the mind, should also be concerned with the education of those young minds. However, the task of genetic epistemology is to develop a theory, supported by elaborate quantitative research, which is in some essential way directly related (by means of relevant independent variables) to the individual and social welfare of all children, and therefore of all mankind. To accept any lesser goal would be academically and socially irresponsible and intellectually irrelevant.

Acknowledgments

The accomplishment of a text of this nature is a long and somewhat complex task, involving many people in varying degrees. The authors are grateful to those persons closest to them personally through this time. For Prof. McGuire, this was Prof. Sally McGuire, his wife; for Prof. Rowland, acknowledgment is due to a constellation of friends and colleagues, of whom the most frequent critic was Mr. E. Gene Patterson (National Association for Retarded Children) who worked with him throughout the production.

Then we would like to acknowledge our profound admiration for the scholars discussed in this volume. Some are our friends, such as Profs. Bruner, Hunt, and Sigel, but all are our teachers and we are deeply grateful.

The authors are also grateful to the reviewers, whose careful analyses gave additional depth to the volume. These were Prof. Irving E. Sigel, SUNY at Buffalo, Prof. Beth Stephens and her associates at Temple University, Prof. Gil Voyat, Yeshiva, and

Prof. Luiz F. S. Natalicio, formerly of the University of Texas at Austin. A particular debt is owed to Mr. Brent L. Hickman, graduate student in developmental education at New York University, for his assistance during the later phases of preparation.

Of special personal and professional importance to both authors were Prof. William A. Hunt and his wife, Diana, editors of *Psychology in the Schools*, in which most of the articles were originally published. Our appreciation also goes to Prof. Jack I. Bardon of the *Journal of School Psychology* and to Prof. Frederick A. Rodgers, editor of "Research in Review" of *Educational Leadership*. These journals worked closely with us and for their assistance, we are also grateful.

Finally, we would be seriously remiss if our gratitude to Mr. Edward E. Lugenbeel and Mrs. October Graham of Prentice-Hall, Inc. were not expressed. Their concern for quality, their expertise, and their support during the production process to a large extent made this book possible.

G. Thomas Rowland
J. Carson McGuire

The Mind of Man

*some views
and a theory of
cognitive development*

Cognitive Development in Children: A Structure-Process System

This brief chapter summarizes our theoretical study of the *Structure-Process* approach to cognition, initially proposed at Houston (Rowland, 1967), and subsequently elaborated on by Rowland and McGuire (1968c) and Frost and Rowland (1968a,b). The model for cognitive development, which is the result of a serious attempt to bridge the gap between the psychological laboratory and the classroom curriculum, has been introduced through the Southwest Educational Development Laboratories into school systems in Texas, Louisiana and Arkansas. In addition the approach serves as the theoretical frame of reference for the *Elaborative Language Series II* (Frost and Rowland, 1968b) (Educational Media Laboratories, Austin, Texas). Further research to clarify the Structure-Process system in regard to early childhood education was proposed to the Research and Development Center for Teacher Education at The University of Texas at Austin.

Within our theoretical framework, the cognitive structures are assumed to be multidimensional and interdependent hierarchies, by means of which the human organism emits observable be-

Adapted from a paper presented at XI Interamerican Congress of Psychology; *Sociedad Interamericana Psicologia and Texas Psychological Association,* Mexico, D. F., December, 1967.

haviors which are evidence of the development of "intelligence." Cognition is defined as *knowledge-ordering behavior* which integrates a multitude of sensory, imaginal, and symbolic cues, that is, which sorts information from error in a cybernetic-like process. Consequently, cognition provides an internal scheme to guide motoric, ideational and verbal behaviors. The theory proposes that intelligence is the ability to adapt to an increasingly complex and sophisticated internal or external environmental situation.

Cognitive structures and the processes involved in the emission of intelligent behavior are held to be highly correlated with the neurophysiology of the human brain. The cognitive processes are dually defined. First, *innovation* is understood to be the human being's adaptive behavior resulting from experience (particularly educational). Cognitive innovations are responsive to the intellectual functions of *accommodation* and *assimilation*. The innovative processes are the basis for the second definition, *radical change* in the cognitive structure resulting from a summation of innovations.

Cognitive development is held to be dependent upon specific neural processes, and to be subject to conditioning and manipulation. Such development is believed to be most clearly demonstrated in the individual's verbal behavior, which is subject to certain consistent and sequential patterns. These patterns of cognitive development are seen as necessarily sequential, but in no way is time *per se* a variable of significance in the developmental sequence.

A literate control of language is primary and sufficient evidence of the development of adequate cognitive structures in the educational encounters of children. Without such evidence, education as cognitive development cannot be presumed to have taken place. Only to the degree that such behaviors are forthcoming may the individual's intelligent behaviors be assessed. If we assume that literacy (functionally effective behavior) is the evidence of education, then certain intervening procedures can be planned and the necessary assessments can be proposed.

Accepting the development of intelligent behavior appropriate to the student at a particular time and stage of development as the ultimate goal of educational intervention allows a redefinition of the educational objective. Literate control of the primary linguistic skills is identified as evidence of cognitive development, and the most basic of these skills appear to be *the primary linguistic skills* of speaking and listening. These skills, largely ignored in the educational encounters of very young children, are essential for continued progress on the continuum of cognitive development.

The *secondary linguistic skills* of reading and writing are dependent upon mastery of the primary skills, and may assume a high degree of autonomy from the primary skills only after intensive intervention into the developmental sequence.

The critical role of verbal behavior in the development of intelligent behavior is most cogently discussed by A. R. Luria (1959), who holds that speech not only has semantic and syntactic functions, but also has a pragmatic or directive function. This function is seen to be demonstrated by the control words have over the activity of children and in the way children use words to control the behavior of others. It is later in the developmental sequence that verbal behavior is internalized to control the organism's own behaviors.

It should be noted that Piaget (1947) does not specify verbal control as a stage in the development of intelligent behavior, but identifies it as a particular and unique type of intelligence. Brunerians would hold that this development occurs during the enactive stage (Bruner, 1964a). In all instances, however, the sequential development concept seems to support a notion of a system of hierarchies of cognition, the development of which cannot be linked to time. It is rather the necessary and invariant sequence which is important in the development of intelligent behavior. It is unfortunate that educators have few if any meaningful measures of concept development (Stott and Ball, 1965), especially in the area of early childhood education; as a result, the school tends to operate without sufficient data regarding the minds of the children at the preschool and primary level.

In order to make some realistic appraisal of the child's linguistic development at the beginning of his educational encounter, the educational psychologist needs to recognize that literacy begins to develop from birth, and for most children follows a general continuous pattern which eventually facilitates their mastery of the language requirements of society. Discontinuity in the developmental continuum may well result in failure to achieve mastery of these requirements.

When a child's development has progressed appropriately, and the expectations of the educational institution and those of the home are consistent, continuity exists. If such is the case, the school will not represent a developmental break or radical change, but will allow for smooth interaction by the student. Conceptual retardation in any form reveals discontinuity or radical departure from the appropriate developmental sequence. For the child for whom school is a discontinuous affair, the experience may be unintelligible and over-demanding. It may happen that such a child will attempt to adopt the expectations and values of the school, and become an outcast in the home environment (Rowland and McGuire, 1967).

Learning undoubtedly begins with birth as one of the encounters between the organism and his environments. There is substantial evidence that learning may begin prenatally (Grier, Counter, and Shearer, 1967). Learning in the postnatal environment may initially be passive, in which case the child receives stimulation from the environment, particularly from other human beings with whom he interacts. The child learns to manipulate and modify the environment, and to adapt his own behavior to gain satisfaction of his needs. At this point, literacy is crude and primitive, most often involving gross physical activity.

A differentiated and flexible language system ordinarily develops in a sequential manner, with opportunities for trial-and-error provided by the family. If the developing child is insufficiently exposed to communicative interaction in the preschool years, his skills will most certainly be underdeveloped. Linguistic development during the early stages will depend largely upon verbal imitation, and will most often be monolingual.

In approaching cognitive development from a structure-process frame of reference, a practical application of the theory structures oral language development around observation, experimentation, description, classification, and prediction. Language development is held to proceed through four stages, which are not time-bound. The stages proceed from labeling, to comparative, to creative and ultimately to elaborative language (Frost and Rowland, 1968b). This final stage represents a level of cognitive development which may not always be achieved in all cultures.

Continuity in language development can exist for most children as reading instruction begins and as readiness activities proceed on the basis of the educator's diagnosis of individual need. A program of continuity in language development rests on the assumptions that:

1. Oral language is the basis for all other literacy skills.
2. Reading and other communicative skills are inextricably interrelated.
3. The basic vocabulary of a child is that which he has, understands, and uses.
4. Reading, like language in general, is presently regarded as more psychological than logical, attaining power and function through meaningful interaction with people, events, and objects; not from external rules and restrictions.
5. Fluency in reading is gained through reading appropriate materials and by reinforcing success.
6. Successful reading is largely dependent upon the attainment of a proper match between instruction, materials, and the development of the child.

A common error made by schools is the misunderstanding and disregard of the conceptual structure of disadvantaged children. The problem exists for all children, but is more acute among the socially disadvantaged and very young because of insufficient experiences. Consequently the attainment of a proper match between cognitive structure, educational materials, and teacher behavior becomes more urgent. The cycle of progressive retardation will be broken only if teachers and administrators develop programs for continuity in subject mastery.

The Structure-Process system of cognition is brought into meaningful focus when we realize that literacy and cognition are essential to any subject matter area. The learner is expected to be literate in science, mathematics, social studies, physical education, or any other content which the learner is required to or chooses to master. From this viewpoint the Structure-Process system is applicable to all classrooms and all content material, and is therefore a valid theoretical approach to the development of intelligent behavior and curriculum design and construction (Frost and Rowland, 1968a).

This system forces the educator to define specifically the concepts needed for literacy in the chosen area. In order to make such a necessary definition, the area will demand in-depth understanding. A second major contribution is that, assuming concept mastery on the part of the teacher, the approach facilitates the essential analysis into components for instructional purposes.

The Structure-Process system supplies education with a

logical thought pattern which could easily serve as the map or schema for curriculum design. Literacy is evidence of mastery, and if literacy is not demonstrated then no conceptual development is assumed. All concepts are learned, and the educator is the stimulus or stimulator of most organized or formal learning within the framework of public education.

The role of the teacher is clarified when we consider it to be one of stimulating learning; the teacher's guidance function is but another dimension of this role. Through this concept, a bridge between education and research in the traditional and basic learning theory of psychology is at least partially established, hopefully bringing many of the theoretical laws previously confined to the laboratory to bear upon the urgent empirical and practical needs of the classroom.

In summary, this entire system has been formulated with the view of establishing a working relationship between two closely allied but often dissimilar sciences concerned with human behavior—psychology and education. The role of the first is to understand human behavior, whereas the role of the second is to change human behavior. Like literacy and cognition, one is considered the essential and dependent function of the other, for understanding is the foundation of change.

2

Jean Piaget

For many years, Piaget's work was almost unknown to American psychologists, primarily because of the lack of adequate translations into English of his elegant but difficult French. D. E. Berlyne was one of the earliest to recognize this inadequacy and to take positive action. Berlyne, with Piercy (1950), translated Piaget's *Psychology of Intelligence*. Later, he spent the year 1958–59 with Piaget in Geneva. Berlyne's book, *Structure and Direction in Thinking* (1965), introduces a system of neobehavioristic concepts which permits an integrative conceptualization of thinking. The next chapter includes our evaluations of Berlyne's work: his theories of curiosity as an emotive phenomenon and of learning through discovery, the Geneva research, and recent developments in Russian psychology.

The work of Berlyne and others resulted in a renewed interest in Piaget; consequently, the current influence of Piaget's research and writing can be discerned in the ideas of many important psychologists, including such diverse figures as D. P. Ausubel (1963) and J. S. Bruner (1965). Today most of his works are available to students, many in inexpensive paperback

Adapted from "The Development of Intelligent Behavior I: Jean Piaget," *Psychology in the Schools,* V, No. 1 (1968), 47–52.

editions. One such edition, *Psychology of Intelligence,* has been reviewed by Carson McGuire (1967). The review is a highly condensed form of the essence of Piaget's work, written in technical terms. In addition, McGuire's paper further amplifies what is said below.

Piaget views intelligence as a mental *adaptation* to new circumstances, both *accommodation* to stimuli from the environment and modification of the environment by imposing upon it a structure of its own, i.e., *assimilation.* Thus intelligence as adaptation involves an *equilibrium* toward which the cognitive processes tend (the act of "equilibration"). The equilibration is between the action of the organism on the environment and the action of the environment on the organism. We must remember that language is a partial substitute for action. Symbols, particularly those of mathematics (which are free of the deception of imagery), refer to an action which could be realized. When such symbols take the form of internalized actions they may be interpreted as operations of thought, i.e., an action which can be internalized.

Piaget emphasizes that the development of *logical operations* is crucial. At Cornell he said:

> To understand the development of knowledge, we must start with an idea which seems central to me—the idea of an *operation.* Knowledge is not a copy of reality. To know an object, to know an event, is not simply to look at it and make a mental copy, or image of it. To know an object is to act on it. To know is to modify, to transform the object, and to understand the process of this transformation, and as a consequence to understand the way the object is constructed. An operation is thus the essence of knowledge; it is an interiorized action which modifies the object of knowledge. For instance, an operation would consist of joining objects in a class, to construct a classification. Or an operation would consist of ordering, or putting things in a series. Or an operation would consist of counting, or of measuring. In other words, it is a set of actions modifying the object, and enabling the knower to get at the structures of the transformation [Ripple and Rockcastle, 1964, p. 8].

Piaget continued his discussion of operations, identifying an operation as an *interiorized* and *reversible* action. An operation is linked to other operations and is part of cognitive matrix, that is, an operation is never isolated in the exercise of intelligent behavior.

Piaget relates affect and cognition—all behavior "implies an energizer or an 'economy' forming its affective aspect." The interaction with the environment which behavior instigates requires a form or structure to determine the possible circuits between subject and object—the cognitive aspect of behavior (schemata). *Similarly, a perception, sensory-motor learning (i.e., habit), insight and judgment all amount, in one way or another to a structuring of the relations between the environment and the organism.*

Hunt (1961) identifies five main themes which he says dominate Piaget's theoretical formulations:

1. The continual and progressive change in the structure of behavior and thought in the developing child
2. The fixed nature of the order of the stages
3. The invariant functions of *accommodation* (adaptive change to outer circumstances) and of *assimilation* (incorporation of the external into the inner organization with transfer or generalization to new circumstances) that operate in the child's continuous interaction with the environment
4. The relation of thought to action
5. The logical properties of thought processes

Roger Brown (1965) attributes to Piaget some 25 books and 160 articles, identifying the goal of the Geneva program as the discovery of the *successive stages* in the development of intelligence. Much current American research, such as that at the University of Chicago, reported by Fowler (1966) and Kohlberg (1966), lends strong support to the goals of Piaget's approach to the development of intelligence. Although these researchers support his approach, they make an important clarification by insisting that whereas the development stages identified by Piaget are real and the sequence is constant, the "American misinterpretation" which attributes to time *per se* the status of a significant variable is not valid. The prospective teacher should grasp this concept of sequence without regard to time boundaries, in spite of the fact that the Gesellian interpretation of stages with firm upper and lower time limits has long been a powerful force in American psychology. *This maturationist view of fixed intelligence and predetermined development is no longer considered valid.* Stages, as identified by Piaget, appear to occur in a constant, *invariant sequence,* but there are no time boundaries. To support this viewpoint, Smedslund (1961) and Wallach (1963) independently designed a program with the specific intention of accelerating a child's development. Both discovered that acceleration appeared to be successful only if the child was approaching readiness at the time of intensive intervention; otherwise, it had no significant effect. Piaget (1953) predicted that this would be the case, saying, "When adults try to impose mathematical concepts on a child prematurely, his learning is merely verbal; true understanding of them comes only with his mental growth." [1] Because many of the conceptions studied by Piaget seem resistant to change by training, there would seem to be a readiness factor involved. It may be that readiness is related to sensory stimulation and is at least partially a matter of varied stimulation (Brown, 1965; Bruner, 1959). The term *readiness,* like the term *stage,* should not be linked with a rigid concept of time.

Piaget has hypothetized four distinct but chronologically *successive* models of intelligence: (1) sensory motor, (2) pre-operational, (3) concretely operational, and (4) formally operational. Piaget (1947, trans. 1950) believes that imaginal thought begins at the end of the sensory motor stage and facilitates the transition into the pre-operational stage. Bruner (1964a), in what seems to be a related concept concerning the development of intelligence, identifies three stages in the cognitive representation of experience: enactive, iconic, and symbolic.

The Geneva research begins with some aspect of common *adult* knowledge. The method of inquiry is to ask questions, and data are the responses of the children. An example of this first method of research is reported in many of Piaget's early works such as *Judgment and Reasoning in the Child* (1924). Piaget used a different approach primarily with his own children, when he employed as the starting point a *set of performances.* His method was *naturalistic observation of infant behavior with experimental interventions.* This second approach is demonstrated most effectively in *The Origins of Intelligence in Children* (1936).

In later Geneva studies, Piaget began with *systematic adult*

[1] The authors prefer to use the term "development" instead of "growth." Development is understood to include: (1) increase in mass, (2) differentiation of parts, and (3) coordination of parts. For example, Rosenzweig (1966) in his research with rats suggests that feedback from learning experiences influences development.

knowledge, asked questions, and provided materials for manipulation. The resultant data included the *manipulations* and *verbal responses* of the children. In the United States there is an increasing tendency to set up contrived experiences which parallel Piaget's research. For example, Bruner and his associates (Bruner et al., 1966) in their book, dedicated to Piaget, report many experiences designed for children which are directly related to the Geneva program, such as Olson's (pp. 135–53) experiment on the development of conceptual strategies with 95 children. Of striking similarity is the experiment on multiple ordering conducted by Bruner and Kenney (pp. 154–67).

An important understanding brought out in the research and writings of Piaget and others (Brown, 1965; Fowler, 1966; Kohlberg, 1966) is that typically the child's intelligence seems to be *qualitatively different* from adult intelligence. As a result, the child simply does not see nor understand things as an adult would.

Children's answers to questions, which appear to be "incorrect" from an adult point of view, are not considered ignorant, but are regarded as imperfect understandings of various intellectual matters. Much importance is attached to the observation that imperfect responses in a sample of children often are similar. Nevertheless, the *order* in which the stages succeed one another usually is *constant.* The word *usually* should be stressed since a range of cross-cultural research has not yet been carried out. A beginning has been made by Goodnow (1962, p. 1) who worked with children in Hong Kong. She began her research to determine if children from other milieus would produce results similar to those obtained by Piaget and his associates in Geneva. The combinatorial reasoning and conservation of space, weight, and volume tasks used in the Geneva program were administered to approximately 500 European and Chinese boys between the ages of 10 and 13. *Raven's Progressive Matrices* were also administered. The Chinese subjects had almost no formal education, as opposed to an average of several years of schooling for the European boys.

Goodnow summarizes:

Similarities across milieus were more striking than differences: There was an odd difference between the combinatorial task . . . and the conservation tasks . . . ; replication of the Geneva results was fair to good; the differences accruing suggest a need for a closer look at the concept "stability of reasoning" and at the expected interrelationships among various tasks [1962, p. 21].

Piaget believes that the child in the period of sensory-motor intelligence does not have internal representations of the world, even though he acts and perceives. With the development of *imagery*, the most primitive form of *central representation* (the beginning of the second stage), the sensory-motor period ends.

Piaget holds that the adult's aspects of identity and perceptual constancies are learned, e.g., one of the most difficult things to understand about the infant's conception of an object is that he does not realize that it exists independently of himself. For example, during the period of sensory-motor intelligence, when the child is governed by his perceptions (what he touches and hears, and particularly what he traces), he begins to develop the fundamental categories of experience and a conception of causality begins to form. *Piaget is convinced that intelligence develops out of motor activity, not just out of passive observation*—the wider the range of the activity, the more diversified will be the intellectual operations of the developing child.

For the adult, an object has an identity which is preserved through various transformations by perceptual constancies such as size constancy, shape constancy, and color constancy. Other aspects of the object's invariant identity regardless of changes of appearance are dependent on knowledge of certain reversible operations, e.g., the adult does not suppose that a car is a different car because the rear end does not look like the front end. For adults, disappearance from sight does not imply the cessation of existence, and objects are continuous in time and space. An object retains its identity through changes of position and illumination, and exists outside the domain of personal experience; for example, Australia exists even though one may not have been there.

The first signs of imagery are a particular kind of imitation and play. *Deferred imitation* is imitation of an absent model, and Piaget postulates the existence of a *central representation* that guides the performance of the child. *Imagery* is also suggested by representational play, and it is imagery, according to Piaget, which makes the development of highly symbolic language possible.

Preoperational and concretely operational levels of intelligence are essentially two levels of response to a common array of tasks; from an adult point of view the preoperational is the less adequate or more incorrect. Though the preoperational child uses language to identify things, ask questions, issue commands, and assert propositions, he does not distinguish between mental, physical, and social reality. He may believe that anything that moves is alive, such as a cloud, and will probably believe that a plant will feel a pin prick. He may expect to command the inanimate and have it obey. To the preoperational child everything is originally made or created—all things are *artifacts*. The parents, as sources for everything, may serve as *models* who make and create things. Parental figures who are close to their offspring may also become models for the child's spontaneous conception of a diety. He sees his parents as infinitely knowing and powerful as well as eternal. The preoperational child is enslaved to his own viewpoint, completely unaware of other perspectives and thereby unaware of himself as a viewer. Things are just the way they are. They are unquestioned. This *egocentricism* is reflected in the child's difficulty in explaining anything verbally to another person. The egocentric child assumes that his listener understands everything in advance. It appears that the egocentric child is moving out of his egocentricism in the later development of the preoperational stage. And, in the same way that the onset of imagery begins this period, the movement away from egocentricism seems to end it, as the child begins to function intellectually in the concretely operational stage.

The intelligence of the older child, who has accumulated learning experiences, is more similar to that of the adult in its separation of the mental and physical world. The older child grasps the points of view of others as well as relational concepts which tie objects and ideas together. The preoperational child has begun to develop the perception of constancies (space, time, size, shape, color) necessary for survival. A child who cannot tell the difference perceptually between an oncoming auto a block away and one ten feet away is not likely to survive. To survive in a world of moving vehicles, he has to learn to understand some of the underlying invariance behind the world of shifting appearances. Yet, to a large degree, he is still controlled by perception. Piaget believes that the preoperational child, despite his limitations from an adult viewpoint, operates with an intelligence that is of a different order from that of the concretely operational child, who depends less on perception.

This dependence on perception often leads the preoperational child to focus upon a single dimension of a problem. He is unable to *decenter,* i.e., he is able to recognize a view which he has just experienced, but not able to pick out views he might experience from different positions. The preoperational child cannot treat relations as left or right, before or behind. Compared to his elders, he is lacking in the ability to discern operations—the central events which, unlike the image, do not imitate perception. Piaget believes that cognitive *operations are derived from overt operations*—interaction between the organism and environments (McGuire, 1967).

The *concretely operational* child can deal to a degree with potentiality as well as actuality, an ability which is limited in the preoperational child. The *formally operational* child approaches what is to Piaget the highest level of intelligence: the ability to represent, in advance of the actual problem, a full set of possibilities. The consequences of formally operational intelligence are identified as "characteristics of adolescence" by Brown (1965), who attributes to Piaget and Inhelder the belief that the reformism of the adolescent is a temporary return to egocentricism. Inhelder and Piaget (1958) and Brown (1965, p. 236), identify as "cultural variation" the phenomenon of primitive societies in which *no one attains formally operational intelligence.*

Piaget finds evidence for his theories in the study of games and the rules for games as children play them, since a child's understanding of rules appears to reflect the level of development of his intelligence. The child who plays egocentrically holds the rules to be inviolable, and may feel that they have always existed. In a transitional stage (late concrete operations) boys begin to play elaborately articulated social games. His observations of this stage caused Piaget to ridicule educators who think children of this age are not capable of learning abstract subject matter. For this level of intelligence, a rule can be changed if consensus of the participants is obtained.

Another aspect of the Geneva studies (Piaget, 1932), the *development of moral conceptions,* begins with the understanding that adults judge naughtiness or wickedness on a *basis of intentions,* and can make independent judgments of seriousness as opposed or related to wickedness. In the preoperational child, however, most often naughtiness was judged in terms of perceived *objective damage;* on the other hand, older children judged naughtiness by the intentions of the offender. Similarly, studies of the child's conception of a lie seem to reflect the level of intellectual development. Young children said a lie was "naughty words" while older children believed a lie to be a statement not in accord with fact. For the young child, reprehensibility is proportional to the variance between the falsehood and truth. Too great a departure became a joke rather than a lie, because no one would believe it. Much older children simply saw a lie as an untruth with the intent to deceive. Piaget asserts that in these developmental sequences a child's morality becomes increasingly inward, a process which Brown calls "enculturation" (1965, p. 211).

Hierarchies and Interaction

The most basic and important views of the development of intelligent behavior which have evolved from the Geneva research of Jean Piaget and his associates center around two concepts. First is the notion of a *vertical hierarchy of operations;* and second, that of *a continuous interaction* between the organism and its environments. These are important concepts in the emergent view of intelligent behavior; the developing synthesis, however, stems from many sources, such as studies of human problem solving, neuropsychology, and computer simulation of intelligent behavior. In general, intelligent behavior is defined as a capacity for problem solving based on a unique set of concepts and strategies organized in hierarchical fashion as a result of continuous interaction with the genetic, internal, and external environments.

Piaget has provided psychology with detailed, longitudinal observations of children developing intelligent behavior. His early insights were the result of studies of the development of the child's judgment and reasoning (1923), conception of the world (1926), and physical causality (1927), followed by the development of language and thought (1932) and moral judgment (1932). The detailed observations of his own children (daughters born in 1925 and 1927, and a son born in 1931) prompted Piaget to enter a different phase of research (1936, 1937). He says of his observations of his children, in cooperation with his wife, that the main benefit was that he "learned in the most direct way how intellectual operations are prepared by sensory-motor action, even before the appearance of language" (1952, p. 249). In these familiar studies, Piaget manipulated the objects and used the results to deduce many of his ideas about the nature of a *central process*.

Piaget holds *seriation* to be the basic operational structure, since numbers do not exist except in series, which constitute the cognitive structure of mathematics. He says:

> These operational structures are what seem to me to constitute the basis of knowledge, the natural psychological reality, in terms of which we must understand the development of knowledge. And the central problem of development is to understand the formation, elaboration, organization, and functioning of these structures [Ripple and Rockcastle, 1964, p. 9].

Piaget identifies (pp. 10–14) four main factors which he feels explains the development of operational structures: (1) *maturation* as a continuation of embryogenesis (Piaget defines embryogenesis as the development of the body, the nervous system, and mental functions), (2) the role of *experience* (the effects of interaction with the environments), (3) *social transmission,* for example, language and education, and (4) *equilibration* or self-regulation.

MATURATION

Piaget questions the Gesellian hypothesis that the development of intelligent behavior is a reflection of an interior maturation of the nervous system. Piaget came to disagree with the view that maturation is an explanatory factor as a result of his very early studies of *limnae stagnalis* (mollusks), which touched on the relation of hereditary structure to the environment. Piaget used these studies as the basis for his doctoral studies in biology, although as he reports in an autobiographical article (1952), his interests had already focused on psychology.

The normative approach to human development postulated by Arnold Gesell is fundamentally in error in its reliance on maturation as the factor of principal concern, for as Piaget says the average age at which stages of development occur varies from society to society (Ripple and Rockcastle, 1964, p. 10). According to Piaget, only the order of the stages is invariant; thus he calls them the *sequential stages of human development*.

To illustrate this point, Piaget refers to research studies of Bushmen and Iranians, which showed that the sequence of stages was found to be invariant; whereas the chronological ages of the stages varied greatly. The idea of maturation alone is insufficient to explain the development of intelligent behavior, basically because we know very little about the development of the nervous system beyond the first few months of existence.[2]

EXPERIENCE

Piaget holds that experience in itself is insufficient to explain the development of cognitive structures, though he retains a basic role in such development for experience or interaction with the environment. Piaget's explanation of the factor of experience sheds light on much of his other research. He holds that some concepts which appear at the beginning of the stage of concrete operations cannot be explained by referring to the child's earlier interaction with his environments. In support of this contention, Piaget refers to the concept of the *conservation of substance*. As a result of his research, Piaget has postulated that the child possesses the conservation of substance before any other concepts of conservation such as conservation of weight and volume. Piaget says that no experiment, no experience can show the child that there is the same amount of substance.

Piaget also objects to the idea that "experience" is considered a sufficient explanation of the development of cognitive structures, on the grounds that the notion of experience is equivocal. Piaget sees two psychologically different kinds of experience: physical experience and logical-mathematical experience. He places special emphasis on this difference as it relates to the educational encounters of children.

Physical experience consists of acting upon an object and drawing knowledge about the object by abstracting from it. For example, two objects may be compared to discover differences in weight.

---

[2] Additional insight may be obtained by referring to the work of Donald O. Hebb (1949) which is discussed in this book in the section on central process theorists. The interdisciplinary approach of Hebb reflects dramatically theoretical reformulations in one discipline as a result of scientific breakthroughs in a related field.

Scientific American has continually published articles related to this area of interest; for example, see Michael S. Gazzaniga, "The Split Brain in Man" (1967).

Logical-mathematical experience develops when knowledge is not drawn from the objects, but from the actions effected upon the objects. Piaget says, "When one acts upon objects, the objects are indeed there, but there is also the set of actions which modify the objects" (Ripple and Rockcastle, 1964, p. 12). He exemplifies by relating the counting experiences of a child. In order to count, the child placed some objects in a row, counting first in one direction and then in another. Arranging the objects in a circle, he discovered that their number was the same. He reversed the direction of the circle, and again the number was constant. The child discovered a property not of the objects being counted, but of the action of ordering. He had enjoyed a logical-mathematical experience, discovering that sums are independent of order, whereas the order was an action which he introduced on the objects.

Piaget holds that the coordination of actions leads to logical-mathematical structures, and that logic is *not* a derivative of language. *The source of logic is the total coordination of actions;* the logical-mathematical experience is necessary before there can be operations, but unnecessary once the operations have been attained. In this context, Piaget's views of experience as discussed above become more understandable.

ASSIMILATION

Assimilation occurs when the organism exploits the environment, incorporating something from it into the organism. The process of assimilation is held by Piaget to be both physiological and psychological, and is more easily understood with the aid of an analogy from biochemistry: An organism ingests food from the environment and gradually assimilates the nutritional substances into its somatic structure.

Psychological assimilation operates when the learner sees something new in terms of something already known, or when it acts in a new situation as it did in a familiar situation. This is immediately recognizable as the phenomenon of transfer of learning. Assimilation also includes the investment of importance, familiarity, or value in any object or idea.

Assimiliation occurs in terms of centrally organized structures which may distort receptor input according to their own pattern and nature. This may be seen as an explanation of deviant or hostile behavior as a result of misunderstanding. Piaget defines assimilation as "the action of the organism on surrounding objects, insofar as this action depends on previous behavior involving the same or similar objects" (1947, p. 7). In this context, Piaget states that assimilation is the fundamental relation involved in all development and all learning, and he distinguishes between assimilation and association, saying:

> In the stimulus-response schema, the relation between the response and the stimulus is understood to be one of association. In contrast to this, I think that the fundamental relation is one of assimilation. Assimilation is not the same as association. I shall define assimilation as the integration of any sort of reality into a structure, and it is this assimilation which seems to me fundamental in learning, and which seems to me the fundamental relation from the point of view of pedagogical or didactic applications [Ripple and Rockcastle, 1964, p. 18].

ACCOMMODATION

In assimilation the organism acts upon the environment. *Accommodation* is complementary to assimilation; the environment impinges upon the organism and results in the organism's modification of action, thereby affecting the environment as perceived by the organism.

The epigenesis of the mind comes about through the adaptive processes of assimilation and accommodation. The innate response tendencies and reflexes gradually become transformed through differentiations and coordinations into the logical operations of adult intelligence.

Cohen says,

> Accommodation is the process of trying out newly-formed internalization for fit with the data of reality, with modifications made when necessary. Infants' accommodation is a motor act; later, it can be achieved through thought. Intelligent adaptation is an equilibrium or balance of assimilation and accommodation [1966, p. 213].

SOCIAL TRANSMISSION

Piaget identifies the factor of *social transmission* as fundamental but, like experience, insufficient in itself to explain the development of cognitive structures. Social transmission includes linguistic or educational transmission. A child, however, can receive knowledge via social transmission *only if he is able to understand that information.* The learner needs a structure which enables him to assimilate the information which he encounters. Without adequate structures, assimilation is impossible.

EQUILIBRATION

Piaget believes that intellectual operations proceed in terms of structures-of-the-whole, which denote the kinds of equilibrium toward which "evolution in its entirety is striving . . ." (1952a, p. 256). He designates the equilibration factor as fundamental to the development of cognitive structures (Ripple and Rockcastle, 1964, p. 13). Equilibration is seen as basic, since in the process of knowing the learner is active and consequently, when he is faced with an external disturbance, he will react to that disturbance in order to compensate for it. For these reasons Piaget holds that man will tend toward equilibrium, which when defined by active compensation, leads to reversibility.

Reversibility, according to Piaget, means that a transformation made in one direction is compensated for by a transformation in the other; and equilibration as a process of self-regulation (in the cybernetic sense) is a fundamental factor in human development. As such, self-regulation may be understood to be similar to a cybernetic control system by which the organism (continually active) reacts, adjusts, and directs its behavior in response to feedback from the environments. This concept sometimes is seen as a matter of *progressive compensation.*

Other Basic Piagetian Conceptions

Piaget holds that *adaptation* is intelligence, and man is in continuous interaction with his environments (1936, 1947). *Interaction* is ·defined outwardly as adaptive coping behavior, that is, the organism's ability to adapt to the circumstances of the environment and to survive. Inward interaction is defined as organization, that is, the continuing arrangement and rearrangement of concepts in a hierarchical fashion.

The invariant functions of adaptation, *assimilation* and *accommodation* are seen as complementary processes. Accommodation corresponds to the notion of outer adaptation to the environment, while assimilation corresponds to inner organization, and Piaget conceives these processes as common to the physiological *and* psychological domains.

SCHEMES

In Piagetian terms, a scheme is the part of an action that is transferable to the same situation or is generalizable to analogous situations. It is the result of the invariant functions of accommodation and assimilation. Cognitively speaking, a representation for any behavior exists in the cognitive structures, which serve as the basis for future cognitive and behavioral interaction with the environment. A scheme is subject to expansion, that is, it becomes more sophisticated or is restructured as the educative process of the individual proceeds, presumably throughout his lifetime.

Schemes may exist in the simplest behavior sequences as well as in the most complex intellectual behaviors which Piaget calls formal operations. Schemes are cognitive representations of a class of familiar behavior sequences identified in terms of functions and cognitive structure. In cognition, existing schemes subsume or are modified to subsume future cognitive events.

The Invariant Stages in the Development of Intelligent Behavior

SENSORY-MOTOR STAGE

Piaget reviewed the sensory-motor stage of development, saying:

The first (stage) is a sensory-motor, pre-verbal stage, lasting approximately the first 18 months of life. During this stage is developed the practical knowledge which constitutes the substructures of later representational knowledge. An example is the construction of the schema of the permanent object. For an infant, during the first months, an object has no permanence. When it disappears from the perceptual field it no longer exists. No attempt is made to find it again. Later, the infant will try to find it, and he will find it by localizing it spatially. Consequently, along with the construction of the permanent object there comes the construction of practical, or sensory-motor, space. There is similarly the construction of temporal succession, and of elementary sensory-motor causality. In other words, there is a series of structures which are indispensable for the structures of later representational thought [Ripple and Rockcastle, 1964, p. 9].

Piaget made extensive and detailed observations of the evolution of sensory-motor and later periods on Lucienne, Jacqueline, and Laurent, his own children. He reports six stages during the sensory-motor stage (Piaget, 1936, 1937).

First, the child *exercises ready-made schemes*. He then begins *primary circular reactions* which consists of coordinating motor habits and perceptions. Third, *secondary circular reactions* consist of coordinating motor habits and perceptions to form intentional acts. During this stage, the child characteristically develops prehension, and begins to search for vanished objects. The fourth stage is the *coordination of the secondary schemes*, at which time the developing human begins to apply familiar schemata to n∍w situations. This is a period when the child begins to imitate auditory and visual models. *Tertiary circular reactions* involve the discovery of new means through experimentation, and is a time when the child systematically imitates new models. The final period of the sensory-motor stage is the *internalization of sensory-motor schemata*, a time when the child invents new means through mental combinations and begins symbolic imitation.

PRECONCEPTUAL STAGE

Piaget says concerning the preconceptual stage:

In a second stage, we have pre-operational representation—the beginnings of language, of the symbolic function, and therefore of thought, or representation. But at the level of representational thought, there must now be a reconstruction of all that has developed on the sensory-motor level. That is, the sensory-motor actions are not immediately translated into operations. In fact, during all this second period of pre-operational representations, there are as yet no operations. . . . Specifically, there is as yet no conservation which is the psychological criterion of the presence of reversible operations. For example, if we pour liquid from one glass to another of a different shape, the pre-operational child will think there is more in one than in the other. In the absence of operational reversibility, there is no conservation of quantity [Ripple and Rockcastle, 1964, p. 9].

The preoperational stage is the first period for developing the three factors which he identified as essential for the achievement of intelligence at the reflective level (Piaget, 1947). These critical three factors are:

1. An increased *speed* of thought to allow the phases of action to be molded into a whole
2. An increase in *scope* to allow concrete actions to be expanded by symbolic representations
3. A concern for the *mechanisms* by which results from action are obtained so that a knowledge of the nature of solutions may be developed.

The development of intelligent behavior is, as a result of the necessary development of these factors, more than a phase of simple translation of sensory-motor to operations. The pre-conceptual stage see the beginnings of the acquisition of language. Of equal importance, the pre-conceptual stage is the period during which the child begins to develop imagery. Wohlwill (1967) reports that Piaget and his long-time associate, Bärbel Inhelder, have demonstrated that images are imperfect at the pre-conceptual level, holding that the images by which the child represents movements and transformations are a product rather than a cause of mental operations.

Images developed during the pre-conceptual period tend to be simple and static, based directly on perception. To be effective instruments for mediating performance on conceptual tasks, images of movement and transformation depend upon the prior establishment of concrete operations.

During the pre-conceptual period, the child's imitative processes become interiorized. Repeated imitation assists in development of the capability of evoking absent models. Piaget (1945) contends that *the image is interiorized action,* and images, which can run off more rapidly than sensory-motor schemata, serve to speed up the representative process. Images at least partially satisfy the requirement of speed, an essential factor for the achievement of intelligence at the reflective level. Images also serve to aid in the development of the second factor, which is the increasing of the scope of the child's activity.

The term *pre-concept* is used since at this level of the development of intelligent behavior, the highest adaptive level of thought of a child remains close to imagery or symbolic play.

This supports the postulate that the pre-conceptual stage comes about as the child initially projects symbolic schemes onto new objects, whereas symbolic play is the projection of imitative schemes onto new objects. These schemes are borrowed from models which are imitated rather than original with the child. The appropriateness of a model is determined by the degree of match between the model and the already acquired schemes. Boredom results from confrontation with previously assimilated models, or with models so inappropriate as to make accommodation impossible.

CONCRETE OPERATIONS

The period of concrete operations, which begins with the intuitive phase of the development of intelligence, depends on the child's being able to *communicate with language, expand his attention span,* and achieve a degree of *self-regulation* which permits him to receive and follow directions. Theoretically, this stage is principally concerned with the relationship between the structures of intelligent behavior and the operations of symbolic logic.

Piaget says of the concrete operations stage:

> In a third stage the first operations appear, but I call these concrete operations because they operate on objects, and not yet on verbally expressed hypotheses. For example, there are the operations of classification, ordering, the construction of the idea of number, spatial and temporal operations, and all the fundamental operations of elementary logic of classes and relations, of elementary mathematics, of elementary geometry and even of elementary physics [Ripple and Rockcastle, 1964, p. 9].

The important distinction appears to be the limitations imposed upon the development of intelligent behavior by the nature of the child, that is, at this stage of development the child *operates only on existing objects.* For Piaget, operating invariably involves motor activity, and at this stage the child is conceived to be unable to operate on verbally expressed hypotheses. At this level of the development of intelligent behavior, the child is as yet unable to formulate a problem and to consider the range of possible alternative solutions.

The development of autonomous central neural processes cor-

relates with the intuitive phase, a period of growing conceptualization (Piaget, 1949). Piaget holds that the regulations of
intuition appear with repetitions of various kinds of overt actions
which provide the basis for feedback from internalized standards. In this theoretical model, Piaget appears to be utilizing
concepts drawn from cybernetics, an interdisciplinary science
formulated by Wiener (1948) and greatly elaborated by Ashby
(1952). The process of internalizing overt action in the concrete
operation stage, presumably follows a cybernetic pattern of
feedback and is fundamental to the development of the basic
operations: classes, relations, and numbers (Piaget, 1942).

To summarize, Piaget holds that the operations essential to
the development of intelligent behavior come about as the result
of internalizing overt actions (always involving motor activity).

FORMAL OPERATIONS

For Piagetian theorists the development of formal operations
is the ultimate major transformation in the development of
intelligent behavior. Piaget says:

> [I]n the fourth stage, these operations (concrete operations) are
> surpassed as the child reaches the level of what I call formal or
> hypothetic-deductive operations; that is, he can now reason on
> hypotheses, and not only on objects. He constructs new operations,
> operations of propositional logic, and not simply the operations of
> classes, relations, and numbers. He attains new structures which are
> on the one hand combinatorial, corresponding to what mathematics
> call lattices; on the other hand, more complicated group structures.
> At the level of concrete operations, the operations apply within an
> immediate neighborhood: for instance, classification by successive
> inclusions. At the level of the combinatorial, however, the groups
> are much more mobile [Ripple and Rockcastle, 1964, pp. 9–10].

This transformation is initiated during the early school years,
and can be seen most clearly in the school's expectations for the
developing child. For the child at the primary level (usually
first through third grades), the curriculum is generally limited
to the manipulation of concrete objects or developing the ability
to cope with actuality in the environment. As the child enters
the intermediate level (usually fourth through sixth grades), he
is expected to begin working with hypothetical situations and
propositions, which become increasingly abstract at higher levels

of the educational encounter. He eventually will be expected to deal with *the possible* as well as the actual.

For example, in the first grade, mathematics may be introduced by having the child group actual objects, and begin to manipulate those groups as he develops concepts of addition and subtraction. At the sixth-grade level, the youngster is expected to be considerably more erudite. His earlier processes are presumed to have become dominant, and in some instances, at the intermediate level he probably will be introduced to geometry and algebra. This process of increasing the abstract nature of the subject matter is critically dependent upon the learner's having mastered the concrete operations stage, which may be viewed as a readiness preparation for formal operations.

Piaget (1947) states that during the stage of concrete operations, the development of intelligent behavior involves the *decentering* and *reversibility* of thought. The learner who has achieved formal operations can verify a law, follow an argument, and state hypotheses, *disregarding the concrete content*. Issues and principles become important as the adolescent sees "what is" in society, and is able to hypothesize what he believes "ought to be."

Piaget (1947) also states that there are relatively few basic operational structures, on which are based a wide variety of thought skills. The development of formally operational intelligence is held to be a "vertical separation," since the earlier concrete operations are grouped into the new structures. Inhelder and Piaget (1955) elaborated and described the few basic structures of thought which have been vertically separated, holding that there are sixteen binary operations of two-valued propositional logic, which are identified as second-order formal operations.

3

∽∾∾∾∾∾∾∾∾∾∾∾∾∾

Daniel E. Berlyne

∽∾∾∾∾∾∾∾∾∾∾∾∾∾

Daniel E. Berlyne, Professor of Psychology at the University of Toronto, is enjoying a diverse and productive career as a neo-behavioristic (S-R) psychologist whose work is relevant to understanding and maximizing outcomes of the educational encounter. He visited Piaget in Geneva during 1958–59, and the influence of Piaget's ideas (translated into neobehavioristic terms) is evident in Berlyne's recent book (1965). Berlyne was instrumental in the publication of the 1950 English translation of Piaget's book, *La Psychologie de L'Intelligence,* first published in France in 1947.

Berlyne's lucid writings on motivation and the thought processes of human beings are as important as his work with Piaget. This chapter will be concerned with Berlyne's thoughts on curiosity and cognitive conflict as motivational factors in *epistemic* (knowledge-seeking) *behavior.* Four of Berlyne's works (1957, 1963, 1965, 1966) were selected as a "core" to the development of his thought. This interpretation of Berlyne's relation to the development of intelligent behavior can by no means encompass the totality of his work; however, the authors

Adapted from "The Development of Intelligent Behavior II: Daniel E. Berlyne," *Psychology in the Schools,* V, No. 2 (1968), 106–13.

believe that it exposes a significant trend in the psychology of intelligent behavior.

There is considerable overlap in the primary resources, since each revolves around Berlyne's interest in curiosity and human thought processes. To avoid redundancy, we will attempt to bring out the important differences and the relationships among the writings reviewed and interpreted. The reader should realize that although there are four primary sources for this representation, there is only *one subject;* thus, each source provides support for Berlyne's conception of human thought.

Berlyne states in his book, *Structure and Direction in Thinking* (1965), that motivational phenomena generally fall into one or more of three classes:

1. *Activation:* the degree of intensity of motivational factors necessary to keep the organism acting
2. *Direction:* the motivational factors that remove uncertainty left by a stimulus pattern associated with a number of alternative responses, and determine which of the several responses shall be performed
3. *Reinforcement:* the association of a particular stimulus or stimulus pattern to a particular response or response pattern, sometimes conceived in hedonistic—pleasant *vs.* unpleasant— terms and, at other times, within the homeostatic framework as a reduction of drive or tension

Berlyne holds that these three motivational factors may be specific to directive thinking. *Activation* in thinking appears to involve the degree of effort, concentration, and force as related to persistence in thought and ability to ward off distraction and discouragement once a chain of thought processes is initiated. *Direction* in thinking encompasses the selection among alternative kinds of symbolic material at different levels. *Reinforcement* usually means the achievement of a symbolic sequence or pattern regarded by the thinker as sufficient for his needs, and appears to be analogous to goal attainment in motor activity. In thinking, the attainment may take the form of confirmation when a thought pattern resolves cognitive conflict.

The Nature of Exploratory Behavior

In discussing the motivation of exploratory behavior, Berlyne distinguishes two types (1965, p. 244). First, *specific exploration,* aimed at and reinforced by the prolongation or intensification of stimulation from particular sources, appears to be tripped off by an aversive condition (imbalance) called *perceptual curiosity*. Incomplete perception may leave the person with uncertainty, which is reduced by exploratory responses designed to obtain additional information. The second type of motivation, *diversive exploration,* introduces stimulation from any interesting or entertaining sources. Specific and diversive exploration, as we will show later, may be related to directed and autistic thinking, respectively.

Berlyne states that specific exploration can be reinforced only by information that is capable of reducing uncertainty, that is, information coming from the object to which the uncertainty relates. As stated above, specific exploration is postulated to be clearly related to directed thinking, whereas diversive explorations are more closely related to autistic thinking.

The principal determinants of specific exploration are labeled by Berlyne as *collative variables* (1965, pp. 245–47) with empirical referents such as "novelty," "change," "incongruity," and "complexity." To collative variables, Berlyne attributes two specific properties. First, they possess *close links with the concepts of information theory* (see George A. Miller, 1956), and they depend on collation (or comparison) of information from different stimulus elements. Second, *collative variables entail conflict,* which Berlyne defines as the simultaneous instigation of incompatible responses. The nature of conflict is deduced from a knowledge of what stimuli are acting on the organism and the behaviors they evoke individually. As envisioned by Berlyne, conflict is not something a person is or is not involved in. For the alert individual there is a constant degree of conflict which will vary from moment to moment. Conflict is closely related to uncertainty (a core concept of information theory); however, *uncertainty reflects probability of alternate responses,* whereas *conflict also depends on their absolute strengths.*

Berlyne refers to the increasing body of experimental research (both Russian and Western) which leads to the conclusion that exploratory behavior is commonly accompanied by increased *arousal* (1965, pp. 251–55). Arousal appears to be associated with general activating or energizing effects, similar to the long-established concept of drive. *Drive* increases when an organism is subjected to physiological disturbance, and may also be increased by external or internal stimuli which have been regularly paired with such disturbances. Motivating conditions aroused by the peripheral organs of the body may be called "sources of extrinsic motivation," and can actuate exploratory or epistemic behavior. Certain forms of intrinsic motivation also are capable of actuating exploratory or epistemic activity. This cognitive conflict depends primarily on the collative properties of the external environment when it is perceived to be vague, hazy, and relatively incomplete.

Exploratory behavior appears to reduce drive resulting from exceptionally novel, surprising, complex or puzzling stimulus patterns. The act of exploration provides reinforcement, which facilitates the retention of the information obtained, consequently reducing the drive.[1] Such reinforcement would tend to strengthen the individual's inclination to engage in exploratory activity in comparable situations.

Epistemic Curiosity, Arousal, and Conceptual Conflict

On the symbolic level, conflict may be caused by discrepancies and/or inconsistent relations among symbolic processes. Resultant tension can be reduced only by modifying symbolic structures and injecting new information. *A state of strong drive induced by conflict related to the symbolic processes constitutes epistemic curiosity. This condition can be relieved by the acquisition of knowledge,* and therefore leads to epistemic behavior which includes *directed thinking.* Directed thinking is conceived as one of three categories of epistemic behavior, and consists of chains of responses. Since directed thinking is a category of epistemic behavior, the essential tension reduction

---

[1] The reader should be aware of the fact that Berlyne has set out to be a rigorous theorist, maintaining the classical behaviorist notion of drive reduction; thus, he has "externalized" many of the physiological sensations experienced by human beings. The only intrinsic motivation he permits in his system involves "cognitive conflict" which itself ends in the *reduction of tension.* This form of intrinsic motivation, of course, is unlike White's "effectance motivation" linked to the idea of competence (1959) but parallels Hunt's presentation of "the incongruity-dissonance principle" (1960) and Festinger's formulation of "cognitive dissonance" (1957). Hunt likens his conception to Hebb's explanation of fear (1946) defined as withdrawal behavior in response to some unfamiliar object incongruous with prior experiences. During the same year that Hebb published "On the Nature of Fear" (1946), however, Heider had presented his concept of balance and imbalance, and Lewin theorized about "cognitive structure of the life space" (1946, p. 796) with "conflict situations" with "force fields" that influence behavior (1946, pp. 808–9).

is dependent on the obtaining of knowledge. Thus, the relationship between exploratory behavior and epistemic behavior appears obvious.

Arousal also may result from a state of inordinately low stimulation, which we call boredom. Such a state may be relieved by the reception of stimulation from virtually any source, provided that it brings the collative properties of the environment to an optimal level. For example, a person bored by inactivity may become restless and take a walk or watch a television program which really does not interest him. The individual in a state of boredom in a classroom may attempt to escape by daydreaming, doodling, or any number of other activities which may result in conflict with the teacher.

Berlyne's report in the *Journal of Experimental Psychology*, "Conflict and Information-Theory Variables as Determinants of Human Perceptual Curiosity" (1957), relates the results of a series of experiments on the integration of the theory of perception with general behavior theory. Such an integration would appear to require some account of the motivational factors underlying perception. Berlyne holds that the application of the concepts of drive and drive reduction to human exploratory behavior demands some justification. He does not believe that Pavlov's theory of investigatory-reflex explains exploratory behavior, and denies that Guthrie's contiguity theory (that learning occurs when events happen close to one another in time or space) is adequate.

Berlyne's theoretical approach distinguishes perceptual and epistemic curiosity. A principal determinant of epistemic curiosity is the degree of *learned conflict* between the symbolic response aroused by a stimulus situation, which he relates to the information theorist's concept of entropy or uncertainty. A case of such conflict is incongruity-conflict, aroused by incompatible characteristics. Surprise-conflict results when a stimulus pattern fails to confirm an expectation evoked by something which preceded it. Berlyne's theory proposed that *epistemic curiosity,* which instigates a search for knowledge, *increases with the number of previous gratifications (successful outcomes) of the drive in similar situations.*

In order to demonstrate the truth value of the proposition, Berlyne used a tachistoscope, rigged to flash pictures for the subject's examination. Sixteen undergraduates participated. Having been told that the experiments were intended to show

how interesting certain pictures were, each subject pressed a key for as long as he wanted to continue seeing a card, then signalled verbally for a picture change. The number of exposures was taken as an indication of the intensity of the drive aroused. The results tended to confirm Berlyne's theory that curiosity increases with the number of previous opportunities to gratify it in similar circumstances. Thus, the predicted importance of incongruity-conflict was demonstrated.

The experiment was repeated with eighteen five-year-old children. The results were the same except that the *children exhibited more curiosity than the adults.*

In "Motivational Problems Raised by Exploratory and Epistemic Behavior" (1963), Berlyne states that epistemic behavior refers to behavior the function of which is to equip the organism with knowledge, by which he means structures of symbolic responses. Epistemic behavior is divided into three categories: *epistemic observation, consultation,* and *directed thinking.*

Directed thinking consists of chains of symbolic responses, and has a special status among the three kinds of epistemic behavior. Observation and consultation generally occur in conjunction with directed thinking. The resultant knowledge ultimately is utilized through its participation in directed thinking.

Successive trials are means of securing information about the probable consequences of performing similar responses in the future. Stored information will be knowledge if the effects on future behavior are exerted through symbolic responses, representing the external stimulus events from which they derive.

If a response is executed for information only, it can often be fulfilled by a curtailed version, e.g., a lick rather than a taste. If a person must recall and anticipate the consequence of a previously executed response, an even more curtailed version may be satisfactory. Under appropriate circumstances an internal or implicit response (recalling prior experience with the object which does not now need to be present) is all that is required. These are the responses on which directed thinking depends.

Conflict between incompatible symbolic response patterns (beliefs, attitudes, thoughts, ideas) is identified by Berlyne as *conceptual conflict,* which he hypothesizes to be a factor producing epistemic curiosity. Overloading of information-handling capacity may result in conceptual conflict. Berlyne acknowledges that this may be what Bruner called *cognitive strain* (Bruner, Goodnow, and Austin, 1956; Bruner, 1956). (Bruner hypothesized that *one* of the reasons for developing cognitive strategies was to reduce cognitive strain.) Berlyne states that the degree of conceptual conflict is assumed to increase with: (1) the number of competing responses; (2) how nearly equal in strength the competing responses are; (3) the total absolute strength of the competing responses; and (4) the degree of incompatibility between competing responses. He identifies six major types of conceptual conflict:

1. *Doubt,* which is the conflict between tendencies to ascribe and to deny reality to a phenomenon.
2. *Perplexity,* which occurs when there are factors inclining the person toward each of a set of mutually exclusive beliefs.
3. *Contradiction,* which is an intense conflict brought about when the person is influenced by factors that not only favor but imperatively force on him two incompatible beliefs.
4. *Conceptual incongruity,* which is a matter of at least two properties and the reluctance to believe that they can be coupled. In this situation, a person believes that A and B never occur together, yet he is confronted with an object that possesses both (a fish that can walk on dry land).
5. *Confusion,* which is produced by information with unclear implications.
6. *Irrelevance,* which is especially related to human thought processes, for example, when an irrelevant thought is allowed to survive even though it interrupts the thought processes.

Relieving conceptual conflict appears to be a result of directed thinking which has both information-rejecting and information-gathering aspects. The reduction of the tension caused by conceptual conflict appears to come about by the *information provided by directed thinking.* Berlyne (1963) posits four conflict resolving methods: disequalization, swamping, conciliation, and suppression.

Several lines of evidence tend to confirm the theory that conceptual conflict can generate epistemic curiosity and motivate epistemic behavior. Relief of conceptual conflict can provide reinforcement for the learning processes by which knowledge is acquired. Berlyne's experimental work has taken the postulated determinants of conceptual conflict and tested their effects upon epistemic curiosity, measured through verbal reports. He refers to the Russian studies of Morozova (1955) and her associates which were devoted to "interest" in school children. One intriguing criterion studied was the literature which interested children. The books most in demand raised questions, offered chances to guess answers, and required thought on the part of the child. Most of the rejected books were simple purveyors of information.

Discovery Methods

Berlyne comments in his brief section on *discovery methods* in education (1965, pp. 264–69) that most new techniques rely heavily on stimulating independent discovery of facts and development of individual judgment. The student is not viewed as passive or absorbing. His curiosity must be aroused so that he will discover knowledge through his own activities. Berlyne says discovery methods may lead children to assimilate material that previously might have been considered too difficult, and that these methods are primarily manipulation of conceptual conflict.

Current research tends to support the emphasis on discovery learning as a valuable technique with children in the educational encounter. Suchman (1960), for whom discovery learning is equivalent to "inquiry methods," began the classroom experience with a film in which some surprising physical phenomenon was demonstrated. Then the children were encouraged to ask questions, and were directed to discover the most appropriate answer. Zankov (1957) reported observing children who were allowed to examine an object during a lecture, rather than merely listen to the lecture, and the results appear related to discovery learning. Milerian (1960) studied the transfer of skill from operating a lathe to operating milling and drilling machines. When the experience of the subjects proved to be inadequate for the new situation, the researchers compared differences between the new task and the old operation, usually generalizing some practical course of action. Kersh (1958, 1962) studied the advantages of using discovery methods rather than traditional methods in the teaching of math; he concluded that students using discovery methods are more motivated to learn.

Epistemic curiosity is usually stirred up by an experience which contradicts expectations or leaves the student perplexed. Berlyne states that the newer methods are aimed at fostering "understanding," and that there are signs of success. It seems apparent that understanding will tend to eliminate conceptual conflict, will be reinforcing, and will therefore be more readily retained by the student.

In his *Scientific American* article, "Conflict and Arousal" (1966), Berlyne discusses the interest of psychology in certain aspects of behavior which hinge on collative stimulus properties (how novel, surprising, complex, puzzling, or ambiguous a stimulus is). These factors can have a wide variety of motivational effects on behavior, even inducing fear and fright. Moderate amounts of unpredictable change are, however, sought out and welcomed. Cases in point are the types of behavior which we label as *play* or *recreation*. It appears that the nervous systems of higher animals can cope with stimulating environments which challenge capacities.

The collative properties of stimuli (that is, phenomena which accompany their perception) appear to have motivational significance because they give rise to conflict. If the nervous system of the organism were not able to handle collative stimuli, adaptive behavior would soon be impossible. To this end, perceptual and thought processes impose order on the external world by classifying, interrelating, interpreting, and organizing the information coming in through the sense organs (i.e., by placing incoming cues relative to one another). Additional information may be sought through exploratory behavior. Berlyne states that there are indications that learning motivated by curiosity can result in not only particularly rapid and lasting acquisition of knowledge, but also knowledge in which ideas are fruitfully pieced together into coherent structures.

A state of high drive manifests itself in three ways:

1. It activates the organism as a whole.
2. It inclines the organism toward a particular class of behavior.
3. It enhances learning by making the organism sensitive.

Arousal and drive are closely related. An exceptionally high state of arousal impairs performance—an effect which has been attributed to "overmotivation." Berlyne states that different kinds of arousal are controlled by different centers in the brainstem (Lindsley's afferent reticular arousal system, 1957), and that changes in arousal are intimately connected with the reinforcement of learned responses. There appears to be a close relationship between arousal and exploratory activity, usually evoked by novel, complex, or ambiguous stimuli.

Berlyne reasons that if we may conclude that conflict can increase arousal, and if arousal may be equated with drive, then conflict will have to be added to the list of conditions which result in a strong drive. In this line of thought, our view of motivation will be broadened considerably if we are obliged to accept conflicts as an additional source of drive.

Epistemic behavior generally is initiated by a *specific* dissatisfaction. The knowledge necessary to satisfy specific epistemic curiosity must be directly related to the original dissatisfaction. The receipt of information cannot be rewarding without the initial conflict, in view of the fact that the psychological function of information is to reduce conflict.

The extrinsic-intrinsic dichotomy is applicable to epistemic behavior. In the extrinsic case, knowledge is welcomed for the contribution it makes toward the attainment of a practical goal. In the intrinsic case, however, knowledge is satisfying in itself and for its power to reduce conflict. The kind of conflict which underlies *epistemic curiosity is termed conceptual conflict* by Berlyne. Conflicts are inseparable from the essence of thinking, i.e., *processes that would result in overt action if not cut short before motor action begins.* In this conflict there is a factor which initiates the process and another factor which intervenes and prevents its progress. Thus, these factors make contradictory demands on the individual's nervous system, and conflict results.

Conceptual conflict is also implied by the nature of intelligence, which comprehends highly abstract concepts by seeking more detail before making a response. Thought objects become gradually less abstract. Beginning with elements that specify a minimum of properties (dimensions), more and more details are supplied so that the area of vagueness is narrowed. In the language of information theory, as interpreted by Berlyne, thinking begins with a high degree of uncertainty, which is then diminished step by step. At the point of uncertainty (vagueness), there is a high degree of conflict which results in curiosity. Conflict and curiosity are reduced progressively until one line of behavior becomes dominant.

Berlyne states that epistemic curiosity plays a part even when knowledge is sought as a means to a mercenary end. The work of Zajone and Morrissett (1960) is utilized to show how receptiveness to offered information may vary with degree of uncertainty. The more uncertain subjects showed a tendency to change their personal opinions when confronted with the opinion of an "expert." Berlyne's own work (1954) with high school subjects demonstrated that an increase in knowledge may result from an aroused high level of epistemic curiosity. In another experiment, Berlyne (1962) hypothesized that epistemic curiosity would increase the number of alternative responses, and the hypothesis was confirmed.

Berlyne believes that the stress on conceptual conflict may have important implications for teachers. Rather than holding to the Herbartian notion that new materials should correlate with the prior experiences of students (the "apperceptive mass"), Berlyne says that a head-on clash between new material and prior experience may best motivate intellectual inquiry and accomplishment. He also says that there are areas where this may not be applicable.

To support his contention of the pedagogical importance of curiosity, Berlyne cites an experiment conducted by J. S. Bruner (1961) in which children were to pinpoint the location of cities by the physical features and natural resources of a region. The children later compared their hypothesized locations with actual locations. The experiment engendered zeal, interest, and better retention, which Berlyne attributes to epistemic curiosity induced by conflict among the multitude of possible choices and doubts about how near the children's inferences would correspond to actuality.

4

⌘⌘⌘⌘⌘⌘⌘⌘⌘⌘⌘⌘⌘

Robert W. White

⌘⌘⌘⌘⌘⌘⌘⌘⌘⌘⌘⌘⌘

The concept of drive theory motivation, in some form or another, appears to have been adopted by many American teachers. Simply stated, this concept proposes that the stronger the drive imposed by the teacher, the more the student will learn. The psychologist immediately recognizes the formula of Clark L. Hull:

$$_sE_r = {_s}H_r \times D \times K \times V.$$

The formula interpreted verbally means that the student's *excitatory potential* ($_sE_r$), which is taken as evidence of learning, equals a multiplicative relationship of habit ($_sH_r$), drive (D), incentive (K) and stimulus intensity (V). The key word is *multiplicative*. Kenneth W. Spence (Hilgard, 1966, p. 189), who both influenced Hull, and was influenced by him, held that incentive was an additive factor. Many psychologists such as R. W. White (1959) seriously question the role of *drive* as a factor in much of human motivation. White proposes that drive may urge the human organism to satisfy the primary needs such as hunger and thirst, and may result in narrow and highly

Adapted from "The Development of Intelligent Behavior III: Robert W. White," *Psychology in the Schools*, V, No. 3 (1968), 230–39.

specialized learning. Drive, as currently conceived in the behaviorist approach to learning, however, is held to contribute little if anything to *developmental* or *general learning,* the type of learning most appropriately the objective of the behavioral science of education.

We must recognize that White *does not* repudiate drive theory as postulated by Hull or by Freud. He points instead to what he considers important inadequacies in the drive theory, in particular, that it does not explain some observable behavioral phenomena in higher organisms, especially in human beings. He also indicates that there are equivalent inadequacies in other motivational models.

Cofer and Appley do repudiate the drive theory model (1966, p. 837), saying "the drive concept is without utility . . . it is a liability," and discuss White's competence model, explaining that effectance motivation is "the production of environmental change" (p. 615).

White says that the main direction of behavior is not to reduce stimulation, as held by drive theorists and some hedonists (who specify as the goal of behavior the escape from painful or noxious stimulation), but rather it is *to vary the manner in which a stimulus acts on the sense organs.* Such an approach might be used to explain in psychological terms the beauty of music or other esthetic experiences.

Effectance

White argues for a *competence motivation* as well as competence in the sense of an achieved capacity. He holds that competence motivation is not random; rather, it is directed, selective, and persistent because it satisfies an intrinsic need to cope with the environment. This competence motivation is called *effectance* by White, who in his earlier work (1952, pp. 247–48) refers to observations of children revealing playful, manipulative, and exploratory activity done "for the fun of it," which serves a serious biological purpose. White charaterizes the activity which is fun as a "feeling of efficacy—or sense of mastery," and further states that "the biological purpose is clearly the attaining of competence in dealing with the environment."

Effectance motivation is seen as not having consummatory acts; and external stimuli, though important, are secondary.

Satisfaction appears to lie in the arousal and maintaining of activity, which need not be intense, but may simply be what would be called *play*. Such playful activity would be overridden by strong primary drives, although White holds that it occupies the spare waking time between episodes of homeostatic crisis or urgent drive. Much of the time spent by children and adolescents in the educational encounter would appear to be under the influence of effectance motivation, and an understanding of this motivational concept takes on additional importance for the educator if this is true.

In effectance motivation, dealing with the environment means carrying on a continuing transaction which gradually changes the relationship of the organism to the environment. This process can be seen in the play of youngsters. White deplores a molecular analysis of play, saying that such an analysis does not take into account the most essential aspects of this behavior—the *continuity of action and change* between the organism and the environment. Play activity in this context is satisfied by the feeling of efficacy, even without goal achievement. That adult humans indulge in playful behavior without seeking to satisfy a primary goal is demonstrated by the adult's indulging in sex for pleasure, rather than participating merely for biological reproduction. Just as the adult may continue to engage in sex after most of the mysteries and adventures have been explored, so effectance motivation may lead to continuing exploratory behavior when the actual gain in competence may be minimal.

In children, effectance motivation is undifferentiated, with differentiation coming later in life and involving achievement, mastery, construction, and other need satisfactions. This does not mean that play is not serious business in childhood, for it is in play that children discriminate visual and aural patterns and build object concepts, while practicing and learning verbal behavior and verbal control of behavior (Luria, 1959). White holds that play in infants is a time of active learning when effective transactions with the environment and movements toward autonomy are established. He states that the toddling infant has begun to achieve competence as he takes his first steps and begins to effect changes in his surroundings.

In addition to logical and empirical foundations for the concept of effectance motivation, White draws support from biology. He holds that a concept such as competence is necessary for any biologically sound view of human nature, especially when one considers the nature of living systems longitudinally. He relates this to man's survival, since the curious and exploring organism has manipulated his environment, improved on it, and has been able to survive. In man there is an increasing tendency toward autonomy from external stimuli, as he relies more and more on internal motivation. This appears to be true so long as there is any development in man, and *when development ceases, man is dead.* Man's autonomy correlates with his developing competence in his actions and with his surroundings.

Taken in the context of White's statements and reasoning, all civilization is a monument to competence and effectance motivation. The competent man has dealt effectively with his environment to make it more hospitable, and in so doing attempts to develop such conveniences as air conditioning, rapid transportation and communication, efficient housing, hospitals and medicine, and schools to educate his offspring by teaching them to be competent.

Ivan Pavlov

In establishing a biological basis for the concept of competence and effectance motivation, while simultaneously asserting further the inadequacy of drive theory in motivation, White reaches back to a 1908 study by Yerkes and Dodson; quotes Tolman's (1948) suggestion that strong drive narrows the range of cues; finds support from Johnson's (1953) studies of latent learning; and ultimately shows that Bruner, Matter, and Papanek (1955) found high drive to interfere with learning.

It appears that White has either purposely or inadvertantly avoided a very valuable source of support in the work of Ivan P. Pavlov. Pavlov (Hilgard, 1966, pp. 56–57) considered irradiation and concentration of afferent neural stimuli to the cortex to be essential to learning. The cortical involvement in learning functions by concentration of irradiated impulses, and concentration occurs, according to Pavlov, in a situation of *moderate* stimulation. Irradiation which does not result in effective learning occurs during periods of either weak or strong stimulation.

It should be brought out that Pavlov's interpretations of the physiology of the brain subsequently have been shown to be somewhat inadequate, but the point is that most drive theories, to which White objects as inadequate, are generally based on

Pavlov's work as interpreted (or misinterpreted) by American Methodological behaviorists. This primary source of drive theory brought into the structure of conditioned reflex psychology the belief that strong or very weak stimulation *interferes with learning*. It seems that White would have given an important additional dimension to his conceptualizations had he utilized Pavlov's thinking, considering Pavlov's considerable influence on many of the developments in American psychology.

White acknowledges that some *general effectance learning* does occur under intense stimulation, but that under moderate stimulation the human organism can attend to less urgent matters: exploring the environment by manipulation and testing himself and the world about him. In periods of *moderate stimulation* man achieves a broad and skillful ability to cope with his surroundings. The infant learns what effect he has on objects and significant people by interaction, and in periods of non-urgency such as a period of play with parents, effectance learning takes place. There is little doubt that Pavlov would have supported these concepts, just as his theories do when they are applied logically.

Other approaches to motivation may be identified as hedonism and activation, exemplified in the works of David C. McClelland (1951), Paul T. Young (1949), and D. O. Hebb (1949). Hedonism is primarily based on the pleasure-seeking, pain-avoidance principles, which often have referential roots in Freud. The core concept of Young's theory is the belief that positive and negative affective states are necessary to account for arousal, maintenance, and direction of behavior. McClelland and his associates, particularly H. A. Murray, have measured motivation by projective techniques from a Freudian frame of reference. Donald O. Hebb is an *activation* theorist who holds that the problem of motivation is the patterning and direction of behavior. He proposes that simultaneously excited neural cells in the brain constitute *assemblies* of mutually facilitating elements, and that a series of assemblies make up a *phase sequence*, which patterns and directs behavior.

The *homeostatic* approach involves the concept of balance, and is widely held (Freeman, 1948; Lewin, 1951). The term *homeostasis* was introduced by Walter B. Cannon (1932, 1939), extended to behavior by C. P. Richter (1942), and introduced to psychology by J. M. Fletcher (1938, 1942) who felt that most psychological concepts could be subsumed logically and factually within the concept of homeostasis. Charles N. Cofer and Mortimer H. Appley (1966, pp. 364–66) explain that the theory of homeostasis does *not* insist that a disturbed organism return to its prior state of equilibrium, but that stability can be reached by assuming the organism to be an *open system*, enjoying a continuous exchange of energy with the environment. Jean Piaget (1947, translated by Daniel E. Berlyne and Malcom Piercy, 1950; Berlyne, 1965; Brown, 1965) adopts a homeostatic model of adaptive behavior with processes of accommodation and assimilation which return the organism to a state of stability. Piaget defines motivation as *disequilibration*. J. McVicker Hunt (1960, 1961) also adopts a homeostatic model with his concept of *cognitive imbalance*.

The purpose of this positional discussion is to propose that none of the major motivational models are adequate for explaining the interaction of children and adolescents with their teachers. In general, the model of homeostasis appears more closely related to the actual educational encounter, but some additional explanation is needed for adequate understanding. We feel that the laws of *drive theory* as postulated by Hull (1943, 1952), and as defended by his eloquent *ad hoc* logic, are theoretically sound. Nonetheless, drive theory is not always related to the behaviors witnessed in the classroom, nor does the hedonism model suffice to explain many classroom actions. The homeostatic model approaches a complete explanation more closely, especially in conjunction with the useful and meaningful theory proposed by Robert W. White.[1] A structure-process interpretation of motivation (Rowland and Frost, 1970) utilizes the proposals of Prof. White, linking them to *epistemic curiosity* (knowledge seeking behavior) (see Chapter Three, Daniel E. Berlyne) and to Bruner's most recent research (see Chapter Six) in a logical attempt to establish *the drive to know* as the fundamental or basic cognitive drive.

Competence

White's major theoretical concept (his word is "conceptualization"), arises from his conviction that drive theory as formulated by Hull and Freud cannot fully explain operating forces in

---

[1] For further clarification of White's theory, see White (1952, 1959).

human behavior. He states that his purpose is an attempt to conceptualize *competence* which will adequately account for those things in human behavior left unexplained by other theories of motivation.

Competence is defined as referring to an organism's capacity to interact effectively with its environment. Later in his statement, White amplifies the definition saying that competence is the product of what Hebb (1949) calls the *cumulative learning* of a *flexible relationship* between stimulus fields and the effects that can be produced in them by various kinds of action. Flexibility is destroyed by strong drive, from which organisms learn well. The organism, however, tends not to become familiar with its surroundings under the impetus of strong drive. White refers to Piaget's (1952) studies of the development of the concept of substance and permanence of objects, a concept which suggests the abstract ideas of space and causality. White feels that these studies provide evidence for his contention that competence is the result of *gradual learning by interaction of the organism with the environments.*

White states that the attainment of competence cannot adequately be motivated by the energy sources conceptualized in drive theory, since this theory cannot account for man's ability to cope competently with his environment. *This ability of man is not innate, nor instinctual, nor can it be arrived at by maturation.* White states also that in considering the problem of motivation of human behavior, boredom, unpleasantness of monotony, the attraction of novelty, the tendency to vary behavior and the seeking of stimulation and mild excitement must be accounted for.

5

∽∾∿∽∾∿∽∾∿∽∾∿∽∾∿

Jerome S. Bruner

∽∾∿∽∾∿∽∾∿∽∾∿∽∾∿

Prof. Jerome Bruner, in his position at the Center for Cognitive Studies at Harvard University, has become one of the most productive, provocative, and controversial psychologists in the United States. Bruner's work, usually written in conjunction with a number of associates, has earned for him recognition by many psychologists (including the authors) as one of the most creative men currently involved in developmental and educational psychology. His writings have also subjected him to criticism almost vitriolic in nature from several people, such as D. P. Ausubel (1966), who labels Bruner's *Toward a Theory of Instruction* (1966) "a very unimportant book . . . superficial, trivial, and repetitious. . . ." In a similar vein, Ausubel himself is criticized in R. C. Anderson's candid statement in the *Annual Review of Psychology* (1967; p. 158) that "Ausubel's research is seldom cited by others actively engaged in instructional research, very likely because his views are at odds with the prevailing climate of opinion."

In the *Annual Review of Psychology* (1967, pp. 94–95), H. W. Stevenson, in an atheoretical, experiment-oriented chapter on

Adapted from "The Development of Intelligent Behavior IV: Jerome S. Bruner," *Psychology in the Schools*, V, No. 4 (1968), 387–94.

developmental psychology, makes a feeble attempt to encompass Bruner's massive contributions in a two-sentence paragraph, in which he emphasizes the conjectural nature of Bruner's theory building—a point which Bruner himself emphasized at the beginning of his 1964 presidential address to the American Psychological Association, "The growth of mind." Stevenson completely failed to acknowledge the antecedent writing and research which culminated in Bruner's 1964 address.

Bruner not only received strong criticism; he also gave it in a review of George Kelly's *Psychology of Personal Constructs*, in which he referred to Neal Miller, John Dollard, and David Mc-Clelland as "Yale thinkers." He branded their theories as "Watered-down hedonisms and tension reductions" and proposed that reinforcement theory makes man a "pig" (1956). The liveliness of the arguments which revolve around Bruner reveals the spirit of a man who has been a decisive force in the movement of American psychology out of the laboratory, and who has compelled psychologists to consider a view of man based on his interaction with his environments.

The writers disagree with Roger Brown's statement that Piaget, after Freud, has made "the greatest contribution to modern psychology" (1965, p. 197). Bruner is most often compared with Piaget; and sometimes Bruner is accused of restating Piaget's theories (Ausubel, 1966, p. 338). Considering the scope of Bruner's publications since his involvement in the Woods Hole Conference of 1959 on science education, the accusation is not entirely just. Bruner has himself replicated, and has encouraged others to replicate, Piaget's work or tasks; for example, Jacqueline Goodnow's (1962) meaningful application of Piagetian tasks in a cross-cultural study involving Hong Kong children, and the abundance of reports in *Studies in Cognitive Growth* (1966) of Piagetian, semi-Piagetian, and original research done at the Center for Cognitive Studies. The potential critic should remember that one of the essential tasks of scientists is the *replication of research*, and that in order to facilitate the production of such research, scientific information is considered to be in the public domain.

In light of the comparison of Bruner to Piaget, and in view of the authors' opposition to the opinion held by Brown and others, an examination of the contributions of Piaget and Freud to psychology seems relevant and meaningful at this point. The work of Freud was intensely idiosyncratic. Moreover, throughout his career, Freud resisted subjecting his data to statistical interpretation, although this appears to have been less detrimental than would be expected because psychoanalysis, which most completely reflects Freudian concepts, is often used by clinicians in one-to-one therapeutic relationships. With this understanding Freud's contribution is logical. Piaget, on the other hand, has also conducted idiosyncratic research, imposing standardization on neither the experimenter nor the subject. In so doing, Piaget has effectively destroyed any possibility for *exact replication as required by science*. The important difference occurs in the application of Piagetian concepts. A great leap of faith is made in moving from his highly individualized research to an application to a group of children in a classroom. Behavioristic theorists have attempted to incorporate both Piaget and Freud by translation. Freud's concepts have been transposed into behavioristic terms by Sears, Whiting, and their associates (1953) and by Dollard and Miller (1950); while Berlyne (see McGuire and Rowland, 1968) has interpreted Piaget in neobehavioristic terms.

Jerome Bruner has employed concepts similar to those of Piaget; for example, Bruner's notions of enactive, iconic, and symbolic representations (1964a) seem to parallel Piaget's stages of cognitive development. The similarity may reflect Piaget's influence or may simply be testimony that children are much alike in both Geneva and Cambridge. Unlike Piaget, Bruner has attempted to *standardize* his tests and is working toward consistent application with groups involved in the educational encounter, as reported in "The Act of Discovery" (1961a).

Failure to make this critical distinction not only results in unfounded comparisons, but does serious disservice to the logic which should be imbedded in the science of psychology. These are the rational underpinnings of the authors' stand in regard not only to those who compare Bruner and Piaget, but also to those who hastily and unwisely criticize what may be the most meaningful work currently in progress within the field of developmental and educational psychology.

A stand such as the authors have chosen to take, based on thought and experience, is in no way intended to belittle the work of Piaget. For our assessment of this important work, see Chapter 2 of this book.

A complete examination of all of Bruner's writings would be impossible in view of space limitations; however, the following selective examinations, considered chronologically, are proposed as key statements in the evolution of a theory of instruction and meaningful conceptions of the development of intelligent behavior.

A Study of Thinking

A Study of Thinking (1956), with Jacqueline Goodnow and George Austin with an appendix by Roger Brown, was the first major publication of the Center for Cognitive Studies. Bruner here constructs the schema for a theory of cognition, while laying out in detail the plans for and results from some rather extensive research in how the human organism thinks. True to his concern for the human organism, Bruner begins with observable behavior. Man categorizes, and learning and categorization are the most elementary forms of cognition. Two types of categories are distinguished: *identity categories,* which classify a variety of stimuli as *forms of the same thing;* and *equivalence categories,* which classify a variety of stimuli as *the same kind of thing.* Three types of equivalence categories are differentiated, although it is recognized that immediate perceptual cues may alter category types. These are (1) affective (categorization by common affect), (2) functional (categorization by interpolative or extrapolative fulfillment of a specific task requirement), and (3) formal (categorization by specifying intrinsic attribute properties required).

Bruner postulates that the basic processes of categorization are the same for both perceptual and conceptual levels, and that categorization reflects culture, e.g., language, life style, religion or science. He holds that the achievements of categorization are:

1. Reduction of environmental complexity
2. A means for identifying objects in the environment
3. Provision for efficient learning
4. Direction for instrumental activity
5. Permitting the organism to order and relate events

Categorization is anticipatory in nature, and Bruner postulates that *need processes,* which may be either facilitative or disruptive, are at the basis of cognitive activities. Cognitive activity is associated with unique affective states, e.g., tension and frustration; and Bruner holds (p. 17) that the insight experience provides feedback to regulate the flow of problem-solving behavior.

The thinking organism validates categories by (1) recourse to ultimate criterion, (2) a test by consistency, i.e., "fit in context," (3) a test by consensus, i.e., matching categories with significant others with whom the individual has identified, and (4) a test by affective congruence, i.e., a feeling of subjective certainty or even necessity. Bruner says that the attainment of concepts is "the behavior involved in using the discriminable attributes of objects and events as a basis of anticipating their significant identity" (p. 21). The process of finding *predictive defining attributes* which distinguish exemplars from non-exemplars of the class the organism seeks to discriminate is called *formal attainment*. A defining attribute is any discriminable feature of an event which is susceptible to some variation from event to event, and when it is used as the basis for inference it serves as a *signal*. A *criterial attribute* is one used to infer the identity of something, whereas a *defining attribute* is some external, arbitrary determination of a category, e.g., categorization by scientific convention or by legal code.

The extent to which an attribute influences the likelihood of categorization is called the *degree of criteriality*, and ranges from maximum to zero criteriality. A cue may become maximally criterial in response to (1) ecological validity, (2) requirements of the categorizing situation, (3) factors of structure and morphology, (4) cue-preference hierarchies, (5) systematic preference, or (6) linguistic codability.

Three types of categories are defined. *Conjunctive* refers to the joint presence of the appropriate value of several categories. *Disjunctive* relates to a category which is arbitrarily defined to be *x* or *y* or *z*, e.g., a fruit that is either orange or circular or edible. Finally, *relational* categories are defined by a specific relationship between defining attributes, e.g., the income bracket of a man is determined by his actual dollar income *and* his number of dependents.

SELECTION STRATEGIES

Strategies are utilized in the process of concept attainment, and are defined as a *pattern of decisions* in the acquisition, retention, and use of information which serves to meet the objectives of (1) efficient attainment, (2) concept dependability, (3) minimization of strain on inference and memory capacity, and (4) concept certainty. Strategies are held to be flexible in order to meet the demands of the particular situation, rather than fixed.

Basically, selection strategies have three potential benefits for their user: to increase the probability that the encountered instance will contain appropriate information, to decrease the cognitive strain involved in assimilating or remembering information, and to control the degree of risk involved in the decision-making process. Bruner refers to the set of consequences following upon each decision and each outcome as the *payoff matrix* of a decision.

In a Brunerian experiment, the subject is presented with a stimulus card from an array of 81 cards (dubbed "Bruner cards"); each stimulus card varies in the relevant attributes (shape of the figure, number of figures, color of figures, and number of borders). The subject (S) is then told that the examiner (E) has a concept in mind, e.g., two green circles with one border; and it is explained that the cards present before him (put there by E) illustrate the concept. The experiment begins with a positive instance, i.e., S is shown a card that has one relevant attribute correctly illustrated. S then selects another instance to begin attaining the concept. E provides S with positive or negative feedback after each choice.

Bruner distinguishes four *ideal selection strategies:* simultaneous scanning, successive scanning, conservative focusing, and focus gambling.

The *simultaneous scanner* uses each instance encountered as an occasion for deducing which hypotheses are tenable and which have been eliminated. It is a most exacting and difficult strategy. The *successive scanner* tests a single hypothesis at a time, and limits his choices to instances that provide a direct test of the hypothesis. This strategy is held to be much like discontinuity in learning, and it does not regulate the risk involved. A positive aspect is that limited inference is required and memory strain is reduced.

The *conservative focuser*, finding a positive instance to use as a focus, makes a sequence of choices, each eliminating one attribute value of the first focus card, then testing to see if the elimination yields a positive or negative response from E. This strategy demands an orderly array of instances, but each instance encountered is informative, redundancy is eliminated, it is cognitively economical, and risk is reduced, since each instance (positive or negative) will carry some relevant information. The *focus gambler* uses a positive instance as a focus, then changes more than one attribute at a time. This final strategy is either very economical or very expensive, and S's tend to change strategies after a miss.

RECEPTION STRATEGIES

The major freedom of the concept attainer (S) is in the hypothesis he chooses to test, not in the way he tests it: S decides that he will test whether the concept contains two circles, and then begins his testing strategy. His principal freedom was in deciding which attribute value (initial hypothesis) was to be tested. E has no control over this decision. Bruner identifies *wholist* and *partist* reception strategies, and holds that a good reception strategy depends upon the alterability of the initial hypothesis in the face of types of contingencies: (1) positive confirming, (2) positive infirming, (3) negative confirming, and (4) negative infirming.

The *wholist* is a focuser, basing his hypothesis on the entire instance. The *partist* is a scanner, basing his hypothesis on a part of the instance, and this strategy relates to the scanning ideal selection strategies.

The partist strategy makes more demands on memory, often encountering negative infirming instances, and the partist must recall all variables. The wholist, however, obtains a summary of all previous instances with each modification of his initial hypothesis, and tends to encounter more negative confirming instances.

The wholist considers alternatives, and gives them weight according to their anticipated outcome. For example, a student might find himself in the dilemma of choosing between going out on a date or doing his homework. The possibility of his teacher's giving a pop quiz the next day is a most relevant variable. He may consider these alternatives: (1) if he does the homework and there is a quiz, he has a *positive confirming encounter;* (2) if he does the homework and there is no quiz, his date was passed up, but he did learn something, and this would be a *negative confirming encounter;* (3) if he fails to do the homework and there is a quiz, he has a *positive infirming en-*

counter; and (4) finally, if he goes on the date, does not do the homework, and there is no quiz, the student has a *negative infirming instance.*

PRINCIPAL FINDINGS

Bruner and his associates found that it is possible to describe and evaluate strategies to a degree, and that it is possible to measure the effect of relevant conditions upon certain aspects of the categorizing strategies. Bruner found that there are certain general tendencies in the obtaining and using of information, such as the tendency to fall back on cues which have proved useful in the past whether such cues are situationally relevant or not; also, people tend to be unable or unwilling to use information based on negative instances and tend to prefer conjunctive (common element) concepts.

Discovery Learning

Bruner in "The Act of Discovery" (1961a) emphasizes *innate curiosity* as the basic motive for learning as an alternative to the classical conditioning model. Learning is defined as rearranging or transforming evidence in a way that enables a person to go beyond the evidence to gain additional new insights. Discoveries will be made regardless of the environment. Certain environments, however, are more conducive to discovery learning than others. Bruner relates this concept of conducive environment to the classroom, proposing as an alternative to the traditional expository mode of teaching what he calls the *hypothetical mode.* In the traditional mode, discoveries are presented to students predigested for easy consumption and regurgitation on examinations. In the hypothetical mode, however, students and teachers are more equally involved in the learning experience, sharing subject-matter evaluation and decision formation. Bruner states that there are four potential benefits in discovery learning in the hypothetical mode:

1. There is an increment in intellectual potency.
2. Emphasis is placed on intrinsic rewards rather than extrinsic rewards.

3. The student learns how to discover, i.e., the student masters the heuristics of discovery learning.
4. Memory processing is facilitated.

Intellectual potency involves an *expectation of success* and the ability to devise *strategies of approach* to discover patterns or regularities and relationships in the environment. The student's method of approach ranges from a hit-and-miss method which Bruner calls *episodic empiricism,* to the meaningful formation of testable hypotheses, which Bruner calls *cumulative constructionism.* Episodic empiricism is seen as leading to confusion and discouragement, whereas cumulative constructionism leads to mastery of the environment. By placing emphasis on discovery the teacher can guide a student in becoming a constructionist.

Intrinsic motivation is postulated as an alternative to traditional reinforcement theory, which emphasizes extrinsic motivation based on some form of reward. Bruner believes that teaching based on extrinsic reward leads to rote learning, with little or no true understanding and mastery of the environment. On the other hand, self-discovery will tend to increase a student's tendency toward self-reward. Discovery will aid the student in predicting and controlling his environment, and this sense of mastery will function as the basis for comparing additional new ideas. The logical conclusion of this reasoning is that the more a student is able to master his environment, the less he will depend on extrinsic reward.

Learning how to discover emphasizes the processes of organized and planned inquiry and research, aiding the student to discard irrelevant variables and to internalize relevant variables in the process of mastering the environment. Learning how to discover, i.e., mastering the heuristics of inquiry, comes about by practice in actual problem-solving situations.

Bruner holds that storage of knowledge is subordinate to location and information retrieval in the conservation of memory, and that improved memory results from improved organization. He suggests that arousing interest in content and relating it to already known material increases organization. Bruner provides an example from his own research which should have meaning for educators concerning memory conservation. Twelve-year-olds were required to remember pairs of words such as "chair-forest" and "sidewalk-square." The subjects were divided into three groups; the first was simply told to remember the pairs, the second was told to produce a word or idea to give the pair sense, and the third group was given aid in the form of the mediator produced by the second group. Bruner identified three types of mediators. *Generic mediation* ties the two words with some superordinate idea, *thematic mediation* imbeds the terms in a story, and *part-whole mediation* sees one term as a part of the other term. Predictably, the second group which produced its own mediators did best (they recovered up to 95 per cent of the second terms) whereas the other groups recovered less than 50 per cent of the second terms. The subjects did better in recalling the second term when the terms were tied together by the form of mediation they most often used. For example, a child who connected "chair-forest" by generic mediation, saying that both objects were made of wood, performed best when the task of reproduction was presented in a generic framework. From this work, Bruner stated that material which is organized in terms of personal interest and cognitive structures has a higher probability of retrieval upon demand, and that discovering things for themselves seems to make learned materials more accessible in students' memories.

Education

Up to this point, we have traced Bruner's development as a *zeitgeist* leader from an early work in cognition to his interest in discovery learning. The direction of his developing concern with education becomes more clear and specific in the early 1960's, as manifested by his "The Ferment in American Education" (1964c).

In an introduction for a book edited by De Grazie and Sohn (Bruner, 1964c), Bruner holds that American education is in a state of revolution. This period of change is characterized by new conceptions such as the belief that any subject can be taught to anyone at any age in some form that is honest. Bruner's conviction was elaborated in *The Process of Education* (1960), which was the outcome of the 1959 Woods Hole Conference on science education held under the auspices of the National Academy of Science through its Education Committee.

Bruner calls for a fundamental reformulation of science to support education. He demonstrates this reformulation by referring to learning theory which, according to Bruner, is a descriptive discipline that has had little direct influence on the actual conduct of education. Instead, learning theory tends to reflect theoretical debates and is restricted to the occurrence of learning in circumscribed (contrived) situations, with only a tangential concern with the optimal means of causing learning to occur. Bruner holds that the results of such basic research do not preclude applied research which would provide the basis for a theory of instruction complementary to a theory of learning. He feels that until a theory of instruction is developed, scientists concerned with education will be unable to test hypotheses about how to teach a subject effectively, i.e., to give research support to ideas about curriculum.

The Course of Cognitive Growth [1]

Bruner's second major statement of theory, published in the *American Psychologist* (1964a), takes the position that the development of human intellectual functioning is shaped by a series of technological advances in the use of the mind. Intellectual development appears to depend on the mastery of techniques, which are skills transmitted by culture. Thus, Bruner sees the development of intelligent behavior to be the result of *external* and *internal* forces. Bruner posits three systems of information processing by which man constructs models to represent the recurrent features of his complex environments: action, imagery and language.

The first of the information processing models is designated as *enactive representation*, that is, a process by which man represents past events through appropriate motor responses. In this stage of development, segments of the environment are represented in a muscular or physical way.

Iconic representation is the second mode of information processing, and in this stage percepts and images are selectively organized to summarize events. Transformed images stand for events in much the same way that pictures stand for the object pictured.

Symbolic representation is the mode of information processing in which a symbol system (usually language in some form) repre-

---

[1] See footnote, page 10. However, the authors respect Bruner's own choice of terms.

sents the environment by design features which include remoteness and arbitrariness, i.e., a symbol is not required to indicate its referent here and now, nor does it have to resemble its referent. For example, a child's name does not indicate his presence nor does the lexeme "Gene" look like the boy it represents. An additional requirement for symbols is that they must be productive in combinations, i.e., a name means nothing unless it can be used in productive grammar to give meaning that cannot be conveyed either by acts or images.

Bruner states that little is known about the conditions necessary for the development of imagery and iconic modes of information processing, and emphasizes his concern with the transition to symbolic representation. He believes this to be a psychologically complex development, and that it begins at about the age of two, at which time a profound change occurs in the use of language and the child begins combinatorial talking, and possibly, thinking. Language begins to provide the child with a means for transforming experience and after combinatorial word play, to explore possibilities for grammatical productiveness.

Bruner suggests that once a child has internalized language, it becomes possible for him to represent and transform the regularities of his environment with greater scope and power, and that hierarchical classification is one of the most evident properties of the structure of language.

Children develop the ability to go beyond the immediate present, making remote reference to states and constraints not given by the immediate situation. They then seem able to culminate information into a structure which can be operated upon, that is, redundancy in the environment is translated into a model which permits the child to go beyond the impact of immediate experience and to integrate [2] events into a longer sequence. This translation of experience into symbolic form supports the development of a symbolic system of processing environmental events, which system will allow the child to deal with the nonpresent, the remote, qualitative similarity and time while remaining in the present.

Language is an instrument of thought, and the capacity for

[2] By *integration* Bruner is referring to a concept similar to that which G. A. Kelly (1955) calls "anticipation." Man is viewed as a bridge between the past and the future, i.e., he connects two temporal universes and uses his role as a connecting link to anticipate events. Piaget refers to this as a "time-binding" mechanism (1964).

using speech cognitively is not actualized until it is coupled with the technology of language in the cognitive operations of the child. Bruner posits that the models a developing child constructs are governed by syntactical rules.

Bruner states that the development of intelligence depends upon two forms of competence: *representation* of the regularities in the environment and *integration,* which transcends the momentary by developing ways to link the past and future. The development of intelligence is held to move forward in spurts, as the child adopts innovations transmitted in prototypic form by a cultural agent (teacher, parent, or surrogate, and age-mate).

Olson (1966), an associate of Bruner's at Harvard, reports a research study which supports Bruner's theory. Olson worked with children and discovered that at three years of age, children tended to use a *searching strategy* in cognitive tasks, i.e., they instigated a quasi-systematic search, whereas five-year-olds used a *successive scanning strategy* in which they concentrated upon patterns. An *information-selection strategy* was used by seven-year-olds, at which age the child tended to eliminate redundancy in pattern matching and was more likely to achieve the required solution with minimal information. To develop this ultimate strategy, children appear to require the ability to map alternative models of the task and learn some way of abstracting a model from the real thing.

Olson discovered a striking difference between recognition of a pattern and the ability to reproduce it, both of which appear to be prerequisites for the development of higher cognitive strategies. The upward move to information-selection strategies requires that the child be able to deal with the properties or features of several images simultaneously, not merely one image at a time, and in addition, that he be able to construct a hierarchy of "distinctive features," i.e., discriminate between a set of alternatives (be able to conceptualize in the domain of alternatives). Also, it appears that the component skills of mapping, locating, and utilizing information change with the development of the child's powers of representation.

Education as Social Invention

In an address to the Ninth Inter-American Congress of Psychology (1964), Bruner proposed a redefinition of education in terms of four important bases. First, he suggested a reconsideration of the educational encounter based on our increasing understanding of man as a species. Second, he proposed an increase in the understanding of the nature of individual mental development in the light of a clearer understanding of the effects of early environments and the impact of the development of language on thought. The third consideration is a greater understanding of education, which has come about as the result of a decade of intense educational experimentation. Finally, the changing nature of society impels us to redefine how new generations shall be educated.

Bruner sees an important role for the psychologist in each of these bases, for it is the psychologist who sees *what is possible*, and this foresight makes the psychologist a powerful force. Bruner warns that if the psychologist fails to fulfill his role, he will not serve society wisely.

The psychologist's understanding of the nature of human ontogenetic development has produced several important conclusions, none of which has been seriously applied to the redefinition of the aims and conduct of education. First, mental development is not a gradual accretion of knowledge, but more a matter of spurts ahead triggered by the unfolding of certain capacities, which must be matured and nurtured before others can come into being. *The sequence of their appearance is highly constrained,* and they are not clearly linked with age; some environments can slow the sequence down while others speed up development of intelligence. The sequence of stages consists of the enactive, the iconic, and the symbolic.

It seems particularly important to re-emphasize Bruner's point. Time is not the variable of major concern in the development of intelligence. The *invariance of the sequence* of the stages of development is the critical concept, and one should understand that the sequence may be slowed down or speeded up in response to certain environments. Consequently, it would seem that a critical aspect of education is the control and production of a facilitative environment.

A decade of intense research in the educational process has produced several hypotheses. One of them, the theory of readiness, is a mischievous half-truth. One *teaches* readiness; one does not wait for it to appear as a consequence of biological maturation. In these terms, readiness comprises mastery of the simpler skills necessary for higher reading skills.

The second hypothesis is that *cognitive or intellectual mastery is rewarding,* especially when the learner recognizes the cumulative power of learning which enables him to obtain easily what was before unreachable.

Third, contemporary exploration has demonstrated that for the most part education has been and is being conducted in the dark without useable feedback. Evaluation after the job is done has become a substitute for real understanding, i.e., the cure for the disease is discovered after the patient dies. Bruner says that it is more sensible to put evaluation into the educational process *before and during curriculum construction,* guiding the teacher in the choice of materials, the approach, and the manner of setting tasks for the learner. Finally, Bruner calls attention to the complete lack of a *prescriptive theory* on how to proceed with instruction.

Bruner emphasizes first that the principal emphasis in education should be placed on skills in handling, seeing, imagining, and symbolizing. In the educational encounter, self-reward sequences should be established, and if there is any way of adjusting to change, it should include the development of a "meta-language" and "meta-skills" for dealing with *continuity in change,* i.e., we must become literate in transforming the apparent shocks of change into something continuous and cumulative.

If man is to adapt to change, instruction must move toward the sciences of behavior and away from the study of history, for the amount of information is increasing geometrically, and demands a change in our handling of its teaching. The shift to the behavioral sciences will allow the student to understand the *possible* instead of the *achieved,* for it is the behavioral sciences which must be central to the presentation of man, while focus on his achievements diminishes. Finally, Bruner emphasizes that if we are to evolve freely as a species by the use of the instrument of education, we will need to bring far greater physical and intellectual resources to bear in designing our educational system.

Bruner urges the re-entry of psychology into education, for psychology has the tools for exploring the limits of man's perfectibility, and in education, psychology will make its contribution to man's further evolution. Psychology can keep alive the society's sense of what is possible.

The Growth of Mind

In Bruner's 1965 presidential address to the American Psychological Association, he stated that his main interest and channel of development was the relationship between pedagogy and the development of intelligent behavior. He turns to evidence that the full evolution of intelligence depended upon bipedalism and tool using. As human groups stabilized and became more complex, reinvention became increasingly costly and unnecessary. Rather, the mastering of the use of already invented tools became important. Culture transmitted the skills needed for the use of invented techniques, implements and devices, as well as providing amplification systems of action, sense, and thought. Therefore, *a society must convert what is known into a form which can be mastered by beginners,* and the more knowledge man has of development, the more accurate and appropriate that essential conversion will be.

Education is a social institution designed to make this conversion, and the more elementary the level the more serious is the pedagogical aim of forming the intellectual powers of oncoming generations.

Bruner lists five great humanizing forces: tool making, language, social organization, the management of man's prolonged childhood, and man's urge to explain. The great forces cannot be compartmentalized, but interact continuously, especially in the educational encounter. Then Bruner states:

> We clearly do not have a theory of the school that is sufficient to the task of running schools—just as we have no adequate theory of toys or of readiness building or whatever the jargon is for preparing children to do a better job the next round. It only obscures the issue to urge that some day our classical theories of learning will fill the gap. They show no sign of doing so. . . . Our special task as psychologists is to convert skills and knowledge to forms and exercises that fit growing minds—and it is a task ranging from how to keep children free from anxiety and how to translate physics for the very young child into a set of playground maneuvers that, later, the child can turn around upon and convert into a sense of inertial regularities [1965, pp. 1015–16].

Bruner's concern for the field and science of psychology is manifested in his final conjecture that psychology is peculiarly prey to parochialism, and thrives on polygamy with her neighbors, especially the biological sciences, anthropology and sociology. Psychology has made contributions to the science of health, especially mental health. He then asks why psychology has neglected the "growth sciences," of which one is pedagogy.

Pedagogy is defined as the field of inquiry devoted to assisting the development of effective human beings, and the limits of the growth sciences are yet to be drawn.

Toward a Theory of Instruction

Bruner's most recent (1966) statement of theory clarifies many of his ideas; he states, for example, that learning is an intensely personal experience, that children are innately curious and enjoy learning, and that they can learn much more at an earlier age than most educators realize. Bruner believes that these ideas lie behind much of the current curriculum reform effort in the United States.

EDUCATION IS EXPERIENCE REORGANIZED

Bruner postulates that much of the development begins (with the help of teachers as cultural agents) by the person's turning and recording in new forms what has been done and seen before, then going on to new modes of organization with the new products of the recording process. He holds that the heart of educational process consists in providing "aids and dialogues" for translating experience into more powerful systems of notation and ordering.

KNOWING IS A PROCESS, NOT A PRODUCT

A body of knowledge, as represented by the faculties of institutions of higher learning and recorded in authoritative texts, is the result of intense prior intellectual activity. Education's goal is to teach the student to participate in the educative process and thus to establish new knowledge, not to coerce the student to memorize known results. Subject matter is taught to force the student to think for himself and to take part in the process of knowledge-getting, not to produce living libraries of the particular subject.

LEARNING IS ITS OWN REWARD

The will to learn is an intrinsic motive, one that finds both its source and its reward in its own exercise. It becomes problematical only under specialized circumstances like those of a school, where goals are set, students confined, and the path fixed. The problem exists not in learning, but in the fact that the impositions of the school often fail to enlist the natural energies which sustain spontaneous learning. These sustaining energies may be conceived of as *curiosity*, a desire for *competence*, aspiration to *emulate* a model, and a deeply sensed *commitment to the web of social reciprocity*.

External reinforcement may elicit an act and even lead to its repetition, but it does not provide intellectual nourishment in a reliable way so that the learner constructs his own model of his environments and what they may become.

SUBJECT MATTER IS A WAY OF THINKING

Bruner suggests that there is nothing more central to a discipline than its way of thinking, nothing more important in its teaching than providing the learner with opportunities to learn that way of thinking—the forms of connection, the attitudes, hopes, jokes, and frustrations that are part of the heart of the subject matter. The best introduction to a subject is the subject itself, and from the beginning the learner should have the opportunity to solve problems, to conjecture, to quarrel, as is done at the core of the discipline.

TEACHING DISCOVERY

One of the most straightforward ways of stimulating problem solving in educational encounters of the new generation is to encourage it in the professional education of the teacher. Bruner believes that this will come in time, but in the meantime, he suggests that psychologists and administrators can encourage teachers to like problem solving, i.e., discovery learning, by providing them and their students with materials and lessons which permit legitimate discovery and permit the teacher to recognize it. Such materials often create an atmosphere of intellectual excitement by treating things as instances of what might have occurred, rather than simply as what did occur in history.

Bruner isolates a major difficulty, stating that whereas a body of knowledge is given life and direction by the conjectures and dilemmas that brought it into being and sustained its development, students of that body of knowledge often do not share a corresponding sense of conjecture and dilemma. This problem is often overlooked in the bustle and concern over materials and content, but Bruner holds that the difficulty emerges because instruction takes the form of telling-out-of-context-of-action. His solution is to design exercises in conjecture, plan ways of inquiry, and practice problem finding and solving.

THE RESPONSIBILITY OF SCHOLARSHIP

If justice is to be done to his further evolution, man needs a method of transmitting crucial ideas and skills, the acquired characteristics which express and amplify his powers. The task will demand our culture's highest talents. Bruner admits that he would be satisfied if a beginning were made in the form of a recognition that this is the task of learned men and scientists, that discovering how to make something comprehensible to the young is only a continuation of making something comprehensible to ourselves. He would have us realize that understanding and aiding others to understand are simply different points on an intellectual continuum.

6

∽∾∽∾∽∾∽∾∽∾∽∾∽∾∽∾∽∾∾

Bruner on

Cognitive Development:

A Review

∽∾∽∾∽∾∽∾∽∾∽∾∽∾∽∾∾

Bruner's early discussions of discovery learning emphasized the learner's ability to resolve ambiguity in the stimulus field, which depends upon the ability to perceive the stimuli and to process the informational input from the environment. This position has classified Bruner as a functional information-processing theorist, and is closely related to his most recent research into the cognitive processes of infants (1967, 1948b, 1969b), utilizing the T-O-T-E model of Miller, Galanter, and Pribram (Bruner et al., 1969, in Rowland and Anglin, 1971), and Bernstein's (1967) model for a system capable of voluntary activity (Bruner and Bruner, 1968).

Bruner's public discussions of his insights into the infant's struggle to control the environment have taken several forms. In "Up from Helplessness" (1969c), Bruner made a general presentation of his research, which was discussed in more specific detail in "Eye, Hand, and Mind" (1969d). The primary sources for this review and discussion are, however, Bruner's addresses to scholarly meetings in both Europe and the United States, in-

Adapted from "Bruner on Mental Development," *Educational Leadership*, XXVII, No. 8 (1970), 841–45. Copyright © 1970 by the Association for Supervision and Curriculum Development. All rights reserved.

cluding his Heinz Werner Lecture at Clark University (1968c).[1]

In his effort to understand the integration of eye and hand movement in the development of intelligence, Bruner has identified four primary concerns (1969d, p. 224): *intention* (voluntary, self-initiated activity), *skill* (the ability to overcome human awkwardness), *attention* (how the afferent domination of perception and attention alters to become efferently relevant), and *integration* (the "orchestration" of previously separate activities).

The Eye

Bruner believes that human vision guides the development of voluntary hand movements (1969d, pp. 224–27) and the eventual development of hand-to-hand relationships, a uniquely human development which appears to take approximately two years during childhood. Immediately after birth the infant's vision is diffusely distractible, a condition which is followed by a period when human vision is characterized by the obligatory nature of attention, that is, attention seems to be "stuck"; this period is in turn followed by the development of anticipatory and predictive vision. During the first two periods, the human being's attention is directed *outward* to the environment, searching for an object or person on which to fasten. This pattern changes gradually during the development of biphasic attention, which allows the infant to anticipate objects in the environment, that is, the infant moves his attention from one object to another without intermediate drifting. Bruner believes this is what Piaget (1952a) calls a "visual schema" in which objects are related to other objects in the environment.

Bruner places great emphasis on the development of biphasic attention, which he sees as crucial since the human being now *processes* information rather than merely receiving it. This developmental change involves not only the infant's placement of attention, but also the withdrawing and shifting of attention. The development of biphasic attention comes before precise coordination of the hand and eye, that is, before visually guided reaching which involves an *orientative visual matrix*, explaining the individual's appreciation of seen hand movements. This matrix

[1] The authors are grateful to Prof. Bruner for providing the written texts of those addresses, and for his useful indications of relevancies within the total research program conducted at the Center for Cognitive Studies.

includes an understanding of both vision and line of sight as compensated for by eye and head movements. During the period of the development of the orientative visual matrix, the individual is also developing hand-mouth coordination and vision-mouth coordination. Bruner sees the mouth as the terminus of guided reaching activity.

The Hand

In a series of three co-authored papers, Bruner discusses the growth of human manual intelligence as a prerequisite to problem solving (I. Bruner, Simenson, and Lyons; II. Bruner and Watkins; III. Bruner, Kaye, and Lyons, all in Rowland and Anglin, 1971).

I. TAKING POSSESSION OF OBJECTS

In an early study, Jonckheere (Bruner, 1968c) examined the use of simple tools in early childhood. Using children from 18 to 24 months old, Jonckheere required the subjects to retrieve an object beyond arms' reach with a rake. The subjects in this *gestalt*-like study were unsuccessful in accomplishing the task, and Bruner believes they failed because they could not orchestrate the component behaviors into an effective whole; that is, integration of component behaviors appears to be a most difficult requirement in the development of human manual intelligence.

To support his ideas, Bruner and his associates designed a follow-up to the Jonckheere study, which Bruner believed demonstrated a strong relationship between control of skilled behavior and later problem solving, a relationship which Bruner calls programmatic. This programmatic relationship is confirmed by the fact that skilled behavior and problem solving share several features. In particular the two types of behavior share the feature of productivity, that is, a set of component or constituent behaviors with rules for combination are capable of generating a repertory of higher level behaviors. To confirm his interpretation, Bruner used 49 infant subjects from 4 to 18 months old, and presented them with a series of toys. The first presentation was of a single toy, which after being dealt with by the infant was followed by the presentation of a second toy and so on to the third and fourth presentations.

The youngest infant in this study could handle only the initial presentation, but the 6–8 month child demonstrated that simple reach and grasp technique was now embedded in more complex behavior. These children took the second toy, sometimes transferring the object to an empty hand, a behavior which was preceded by the infant's bringing and holding the toy at midline. This gradually changed to an *anticipatory* hand-over wherein the children expected a second presentation, and some crossed the midline to make the transfer either by traversing the midline or by adjusting it by body shift.

The 9–11 month subjects showed another important development. Many could now handle three presentations by depositing one of the toys, but only for a short time. At 12 months, many of the children anticipated additional presentations and were adept at storage for this eventuality. In addition at this age, many had developed to the point that there was a significant reserve of prehensibility which allowed them to accept another toy in the same hand. These older children also used other people, such as their mother, as storage agents.

For heuristic purposes, Bruner uses an information-processing model, and interprets the initial failures of his subjects as uncontrolled activity. Bruner sees the infant's eventual successes with the environment as the first steps toward voluntary and unitary action; that is, effortless subroutines are developed which are in turn incorporated into more complex acts. Bruner identifies this developmental process as *modularization*, a major process requiring several supporting processes, including the notion of developed behaviors triggering behaviors to be developed, a concept sometimes associated with ethology. Bruner, however, sees this as the progressive incorporation of behavioral modules into programs. For explanatory power, Bruner uses Bernstein's (1967) model including,

1. The individual's *effector activity* regulated according to specific parameters
2. A *control element* which conveys the value of the parameter
3. A *receptor* which perceives the factual value of the behavior in relation to the parameter
4. A *comparator* which perceives discrepancies between the factual and required values
5. An *encoding device* which corrects behaviors
6. A *regulator* which in turn controls the effector activity

Like most IP models, this is an open-energy, closed-loop system, and learning affects changes in the control, receptor, and comparator mechanisms.

II. THE ACQUISITION OF COMPLEMENTARY TWO-HANDEDNESS

In this study, Bruner and his associates used 40 children to demonstrate the serial ordering of behavioral components into complex behaviors. The task was to slide up a weighted, transparent lid to obtain an object. The youngest infants could not differentiate the task into components (raise the lid, hold, capture, and retrieve the object). The next oldest group could raise the lid, but the act became preemptive and the intention to capture the object was lost, while opening and closing became an object in itself. One-handed success tended to be continued when it occurred, but two-handed reaching was more successful and gradually took over. Two-handedness emerged in two ways: by symmetric cooperation or by a sequential pattern in which one part failed. Both are developmental and appear to be transitional, the first requiring activation of both hands from the beginning and the second being a near-mastery sequence.

Success in two-handedness increased with age, and appears to be a function of controlling interfering activities such as clawing and banging. As the child's control increased there was less likelihood that component acts would become autonomous. Early success allows for the perfection and reinforcement of serial ordering of modularized behaviors, leading to eventual skill mastery. Bruner believes that the first appearance of a constituent activity is *innate,* occurring in a crudely controlled form, and that it is only after the initial appearance that the behavior is consolidated and shaped. He holds that the emergence of a behavioral pattern or strategy seems independent of practice, and does not develop as the result of trial and error; instead, development appears to be in response to environmental demands and events (operative requirements) which require the *intention* of the learner.

Greenfield, an associate of Bruner's, carried out a cross-cultural study of intention using Mexican Indian children. She defines intention as the internal counterpart of an external goal, and states that intention is both structural and motivational, thereby classifying motivation as intrinsic to the structure of intelligent behavior. In addition, Greenfield believes that the environmental specification of goals determines whether or not learning will take place and the type of learning that will occur. This substantiates Bruner's conviction that the exercise of one's own intentions and learning how to implement them is crucial in the development of intelligent behavior (Greenfield, 1969).

III. THE DEVELOPMENT OF DETOUR REACHING

Using 120 infants, Bruner and his associates studied the development of human manipulatory behavior which they believe emerges from indirect or direct detour reaching. The children were divided into three groups with median week ages of 34, 51, and 69. They were held in their mothers' laps and confronted with an specially designed box (36" × 24" × 18") with a visual area of 5½" × 36" which could be obstructed by a transparent or opaque sliding panel. The object of attention was placed behind the panel. A simple activation of the contralateral (farthest from object) hand would allow easy capture of the object. The room was darkened while the interior of the box was lighted so that the object was immediately visible to the child. Predictably, the youngest children had the greatest difficulty capturing the object in most instances. Bruner manipulated the experimental situation by using both the transparent and opaque screens and there were four positions for the object from "open" (no obstruction) to "deep" (5" behind screen). The screen was at the child's midline. One most intriguing observation was the performance differences by age in relation to the type of screen. The youngest children were more successful with an opaque screen than with the transparent screen, a result which Bruner interprets as a reduction of interference in the detour reaching of these infants.

Several behavioral organizations were observed in this study, and Bruner identified a progression of processes: (1) the activation of the hand nearest the goal (ipsilateral hand); (2) the activation of the appropriate hand; the appreciation of spatial demands; (3) the dissociation of line of action from line of sight; (4) the ability to shift behavioral program; and (5) the ability to sequence instrumental behaviors in order to reach the goal. As in the other investigations of manual intelligence, Bruner sees in the development of detour reaching a close relationship between initial skill learning and later problem solving, all dependent upon the individual's integration of constituent behaviors with prerequisites of intention, skill, and attention.

Summary and Conclusions

Bruner, in his latest research, has evolved several concepts which seem most important for psychologists and educators. First, the development, exercise, and implementation of the individual's intentions appear to be crucial to learning. Second, the development of intelligent behavior seems even more to be a matter of sequence. Third, skilled behavior appears not to be an entity in and of itself; rather, it is the result of the integration of constituent behaviors. In addition, Bruner seems to have shown that the control of the stimulus field and the amount of the information available to the learner correlate with the development of certain skills. He has again emphasized the overpowering impact of the environment upon intelligence, a fact which should be of interest to the educator, since in the school, the educator is largely in control of the environment.

This critical function of the educator as the controller of the child's intentions in the learning process cannot be ignored. The importance of this concept is given incremental explanatory power when we understand that the intention or purpose of human cognitive functioning is *to know*. Berlyne (1954, 1957, 1962, 1963, 1965) defined this process as *epistemic curiosity*, or the seeking of the learner to resolve ambiguity in the stimulus field. Epistemic curiosity has been defined (Rowland and Frost, 1970, in press) as the fundamental or innate cognitive drive, a notion which is directly linked to the *structure-process* model for education (Frost and Rowland, 1969). The fullest implications of Bruner's most recent and perhaps most difficult research can be appreciated only through this theoretical interpretation in regard to education, which begins with the origins of life, called *genetic epistemology* as conceived by Piaget (1947). In this context, Bruner's concepts have provided the medium for irrevocably linking developmental psychology and education, a development which Bruner predicted in *Toward a Theory of Instruction* (1966), a small book never fully appreciated for its psychological and educational implications.

The concept of epistemic curiosity or knowledge-seeking behaviors as innate seems to be logically involved in the evolutionary notion of *survival;* for not to learn is to be *cognitively rigid,* an intellectual state at least analogous to physiological death. In this context, *to educate the oncoming generation* is the primary social responsibility of man as an individual and as a culture. This responsibility is sometimes satisfied by the individual's acceptance of an appropriate role; for example, parenthood or professional education. The satisfaction of the social responsibility to educate by a culture often takes the form of providing *access;* for example, the private and public school systems in most societies. In this context, *developmental education* is the true metascience encompassing the study of man and commanding accessibility to *all* sciences in order to ensure the survival of the culture of man. To provide oncoming scholars and social neophytes with second-rate, biased, or distorted data, or to provide unequal access to quality data, or to charge inadequately prepared professional guides ("teachers") with the care of young minds throughout their educational encounters *at any level* would be tantamount to an act of intellectual anthropophasia; and this in turn would be *cultural suicide.* This destructive process is greatly facilitated when the scholars and scientists of a culture ignore the education of children and surrender the critical profession of education to persons of less than the highest standards of competence, professional ethics, and scholarship.

7

❧❧❧❧❧❧❧❧❧❧❧❧❧❧❧

Irving E. Sigel

❧❧❧❧❧❧❧❧❧❧❧❧❧❧❧

Genetic epistemologists, developmental psychologists and early childhood educators have long been concerned with the development of "symbolic," "abstract," or "formal logical" reasoning. In other words, How does a human learner develop the necessary strategies to cope with or adapt to that which is not present in the environment? What are the structures needed and the processes necessary for the learner to develop the skills essential to understanding and knowing that the three-dimensional present can be, and often is, represented in a less than actual manner such as by a photograph, a sketch, a picture, or even by a gesture? Moreover, what is the essential relationship between the development of these abilities, which Irving E. Sigel calls *representational competence* (1968a,b), and future success within the learner's environments?

Sigel, currently of the State University of New York at Buffalo, began the development of his *distancing hypothesis* as a result of inconsistencies observed during early studies (1953, 1954) and later research in cognitive styles among American

Adapted from "The Development of Intelligent Behavior VII: Irving E. Sigel." *Psychology in the Schools* (1970). We would like to express our gratitude to Professor Sigel for his cooperation, especially for providing copies of the two primary sources for this article.

lower-class children (Sigel, Anderson, and Shapiro, 1966; Sigel and McBane, 1967).

In his earlier study, Sigel came to the conclusion that the mode of representation did not significantly influence the classification abilities of lower-middle-class boys in tasks which demanded that the learner sort familiar items. In these studies, the subjects were required to sort pictures, words, and three-dimensional objects, and the mode of representation was not statistically significant. Sigel concluded that meaning transcended representation, and therefore when a set of stimuli representing a familiar object was presented, the mode of representation was irrelevant; that is, the learner was not confused by representational variations. This observed consistency was designated as *representational competence*.

RECENT RESEARCH

In his later studies, with children from the lower-class black population of Detroit, Sigel noted unexplainable differences in similar sorting tasks. The children used in these studies exhibited difficulty in classifying or grouping pictures of objects according to common characteristics. Rather, the children in this new sample tended to form classifications by chaining in thematic ways; for example, given objects, X, Y, and Z, these children might place the three objects in one class, recognizing that both X and Z are related to Y, but not recognizing any necessary or essential relationship between X and Z, as would be predicted for all objects placed in one category.

Initially, Sigel and his associates concluded incorrectly that these categorizing inconsistencies could be attributed to differences in the sample children's capabilities to group or classify. Upon reconsideration, however, the research team arrived at the notion that a picture or photograph does not always reflect what a child knows or understands about the pictured or photographed object. In other words, the mode of representation was not only important to certain children, but was critical to understanding the development of intelligent behavior in these children.

The impact and significance of this observation, and the resulting theoretical and research developments related to the distancing hypothesis, become even clearer when one recalls that many measures of children's abilities, and their consequent "scores" or "quotients," depend almost solely upon the child's understanding of some culturally common meaning attached to a picture, photograph, or sketch.[1] This meant, as was pointed out to the Sigel team by a kindergarten teacher, that the ability to label a picture did not necessarily indicate the child's understanding about the pictured item. It might mean instead, that the child simply did not understand that the picture was a representation of a real or actual object.

This possibility that the mode of representation was indeed one relevant dimension, if not the chief relevant dimension among certain populations of school children, was verified when Sigel and his associates observed that:

> Given three-dimensional objects, the lower-class children did not show any greater difficulty in grouping than middle-class children, but the difference between the use of pictures between the two socioeconomic groups showed considerable difficulty where greater frequencies of non-grouping and non-performance were found. The difference between the use of objects and pictures was also significant for the lower-class child, but not for the middle-class child [1968a, p. 3].

SUPPORTING EVIDENCE

The distancing phenomenon was supported by several other studies. Sigel and Perry (1968) confirmed the difficulty lower-class black children had when required to cope with the not-present, as measured on the Motor Encoding test of the *Illinois Test of Psycholinguistics,* for example, in which the subject is required to demonstrate by gesture the function of a pictured object. Lower-class black children had considerable difficulty with this requirement, but had less difficulty when presented with a three-dimensional version of the object. When given the

---

[1] This seems an especially meaningful aspect of the Sigel findings in view of the recent revival of the controversy regarding the genetic vs. the environmental contributions and influences to the development of intelligent behavior in the *Harvard Educational Review* (Jensen, 1969), which seems little more than a renewed emphasis of a subject previously discussed within American psychology (Garrett, undated). The nature vs. nurture controversy is not new to science in general or to psychology in particular (see E. G. Boring, 1957), and the only defensible position seems to be that although certainly there is no question that the genetic endowment of the individual will influence his limitations for the development of intelligent behavior, this consideration seems irrelevant to the primary social responsibility of education. The fact that the controversy has been revived and reinjected into discussions of early childhood education, especially of children from the culture of poverty, seems most regrettable—even irresponsible.

actual object, these children had almost no difficulty at all. Sigel believes not only that the child in the three-dimensional condition has multiple cues, but also that "he has the gestalt of the object in its spatial locale, its palpability and its consistency with his own active experience with objects" (1968a, p. 4).

In a second study involving doll play situations with a similar sample of children, the tales of the lower-class children tended to be action-oriented and based in reality. The lower-class children tended to use statements of actions and interactions with only minimal reference to past or future, and rarely referred to inner feelings or thoughts. From these observations of the play behavior of lower-class, black children, Sigel reported minimal use of imagery, but a high degree of motoric, action-based play (1968b, p. 10).

Sigel believes that the empirical differences which he and his associates observed (1968a, p. 5), that is, the discrepancies in object-picture classification, the tendencies for difficulties on motor-encoding tasks, and the motoric action level of the play of lower-class children, can be explained by the difficulty of such children in reconstructing reality in symbolic terms.

Sigel does not believe that this type of response discrepancy is limited to lower-class, black children (1968b, pp. 10–13). He refers to Smilansky's (1968) studies of Israeli children from underprivileged backgrounds, who demonstrated similar difficulties. In addition, several studies conducted at Merrill-Palmer Institute, by Sigel, J. P. Jackson and their associates, supported the contention that even middle-class children found difficulty in building and utilizing logical categories; that is, three-dimensional objects were more readily classified and recalled than their two-dimensional counterparts.

Sigel also sees support for his theoretical hypothesis in Hudson's (1967) report of perceptual difficulties among Bantu tribesmen in South Africa, when they are confronted with pictures as representations of reality. Additional support is drawn from the studies of the influence of ecological setting on the susceptibility to illusions (Segall, Campbell, and Herskovitz, 1966).

Representation

Sigel was unable to adequately account for representational incompetence within existing theories, though he considered the writings of Piaget, Bruner, and Werner and Kaplan.

PIAGET

The primary source for Piaget's consideration of representation is *Play, Dreams and Imitation in Childhood* (1945, pp. 273–74), and it should be recalled that the theory which has come to be identified with Piaget's name was developed almost solely upon observations of middle-class Swiss children, a group which may be considered a unique and limited sample. In addition, Piaget's tendency is to use terms in distinct frames of reference, with shades of meaning reserved to Piaget and his close associates.

Piaget holds that representation is "characterized by the fact that it goes beyond the present, extending the field of adaptation in space and time. In other words, it evokes what lies outside the immediate perceptual active field" (1962, p. 273). Piaget conceives of representation as a union of recall and thought, and holds that language is the primary factor in the development of representational competence. Language, on the other hand, demands an accessible system of "signifiers" which allow recall, and on this basis Piaget believes that the thought of a child may often be more symbolic than that of an adult. He says:

> [T]his "signifier," common to all representation, is the product of an accomodation that is continued as imitation, and hence as images or interiorised imitations. Conversely, the "signified" is the product of assimilation, which, by integrating the object in earlier schemas, thereby provides it with a meaning. It follows that representation involves a double interplay of assimilations and accomodations, present and past, tending towards equilibrium. This process is of necessity a slow one, . . . As long as equilibrium has not been achieved, either there is primacy of accommodation, resulting in representative imitation, or there is a primacy of assimilation, resulting in symbolic play [1962, p. 273].

It seems clear that in Piaget's context, the representational incompetence which Sigel and his associates identified would be understood as an evolutionary process in which equilibrium is not achieved; which process could in turn be identified by the child's inability to categorize and utilize two-dimensional representations of three-dimensional objects. This phenomenon does not seem to be restricted to any particular social class or culture or age, but rather appears to be an individualistic phenomenon directly related to the experiential background of the particular subject in the context of his unique cultural limitations or enrichments.

BRUNER

Bruner, who has consistently insisted on the active nature of intelligent behavior, sees representation in a somewhat different light than does Piaget. For Bruner, representation is the individual's set of consistent behaviors, which he feels are directly related to the overt functioning of intelligent behavior. Some would call these behaviors *stages;* Bruner, however, has identified three modes of representation: enactive, ikonic, and symbolic, each of which is identifiable by the degree of abstractness within each consistent representation (1960, 1964a).

Within the Brunerian context, classification of stimuli, or the building of categories, is the most essential of all cognitive acts (Bruner, Goodnow, and Austin, 1959), and therefore representational competence would be an essential consideration. The well-known "Bruner cards" were used in a two-dimensional manner, with relevant and irrelevant aspects, to identify four cognitive styles, but there was no attempt to take the step from the three-dimensional, present object classification task to the two-dimensional, not-present representation of an object classification task which concerned Sigel and his associates. Many other studies of the learner's abilities to utilize relevant and irrelevant dimensions in cognitive development, such as that of Kendler (1963), have also not taken into account the representational competence factor.

WERNER AND KAPLAN

For Werner and Kaplan (1963), representation is the reconstruction of the world of objects, and such representations serve the learner as guides for action. In this context, a symbol is a fusion of form and meaning, which designates a medical pattern referring to content. Symbols, in turn, serve the function of representation, and are distinguished from signs, signals, or things.

SIGEL

The use of the term *representation* varies among theorists, reflecting the particular emphasis or meaning of the individual formulating and/or using the term. Sigel clarifies his meaning:

[L]et it be made clear that *re-presenting* is the process of reconstruction of reality which can be *external* or *internal,* while *representation* is the product or outcome of re-presenting. Objects can be represented by pictures or symbols, photographs, drawings, art forms, etc. Internal representations take the form of imagery or schematization of reality in various degrees. Representations contain at minimum fractional elements of the reality depicted. Thus, representations can vary in content [1968b, p. 15].

He continues, "Representational competence refers to the individual's capability to respond appropriately to external representations, to behave in terms of internal referents, to reconstruct nonpresent reality." Therefore, representations are nonpalpable re-presentations, external to objects and events, and depictions of reality. Sigel believes that it is essential for a child to learn that representations are modes (distinct from their referents) of depicting objects, persons, or events in three-dimensional reality.

Sigel, unlike Piaget, does not consider language to be necessarily a form of representation (1968b, p. 16), since it can serve to evoke representations and nonrepresentational behaviors, as well as to describe, reconstruct, and anticipate, and often serves the user as a signaling system to others. Sigel makes only one exception in excluding language as a form of representation; he concedes that certain poetic rhythms are representational because they generically fuse in a symbolic way with the referent of the word.

The Distancing Hypothesis

Sigel formally states his theoretical hypothesis thus:

Acquisition of representational competence is hypothesized as a *function* of life experiences which create temporal and/or spatial and/or psychological distance between self and object. *Distancing* is proposed as the concept to denote behaviors or events which separate the child cognitively from the immediate behavioral environment [1968b, p. 16].

His emphasis on behavior refers to demanding that the child react to the nonpresent (future or past) or nonpalpable (abstract language). He thereby clearly places the distancing hypothesis within the domain of science, as defined and understood by most psychologists and other scientists. In addition, Sigel has separated himself, at least on this point, from Piaget by his exclusion of language, a factor upon which Piaget relies heavily for explanatory power. Another way in which he differs from Piaget is that Sigel is obviously concerned with lower-class children and with education, both of which have only been tangentially treated by the Genevan theorists.

Within Sigel's context, distancing stimuli may emanate from persons, objects, or events, and are adaptive in nature insofar as they force the learner to develop and use representations. For Sigel, distance is expressed in terms of the stimuli presented to the child, which force the development and use of representations. Distance may be "close," a concept which he exemplifies by stating that a cutout of a chair is less distant from the child than the word "chair" (p. 17). Distance, therefore, is a function of the overlap of the stimuli with reality; the cutout of the chair has considerably more overlap than the word. This concept of reality overlap is again exemplified by Sigel's contention that the cutout of the chair is possibly less *distal* than a photograph, since the cutout may provide some cues as to dimension or depth or physical form, while the photograph relies almost totally on visual cues. The concept of distance seems most clearly shown in the accompanying diagram.

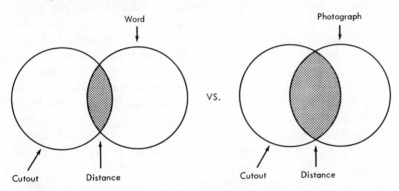

A model for the distance concept.

"Distal" is Sigel's term to refer to the distance within the domain of the stimuli, whereas "distancing" refers to "acts or events which may or may not employ distal stimuli to create the separation or differentiation between self and the physical environment" (p. 17). He has introduced, therefore, two important terms into the consideration of intelligent behavior: *distal*, referring to the nature of the stimulus; and *distancing*, referring to the classes of behaviors which differentiate the individual from the environments. In addition, Sigel holds (p. 18) that distance may be *temporal* (past event vs. present recall) and *spatial* (picture vs. pictured); and that there may be differences in *modality* (name vs. object) and *degree of detail* (sketch vs. photograph vs. object).

Basing his notions on the assumption that man has the capability to create representations of reality, indeed that it is necessary for the child to create such representations in the sequential development of intelligent behavior, Sigel believes that the notion of distancing characterizes the differentiation of the individual from the environments, that is, "the subjective from the objective, the self from others, ideas from actions." Sigel emphasizes two important concepts: that distance development results from the experiential background of the individual, and that this background is composed not only of the unique individual history, but also of the culture from which the individual emerges.

Conclusion

The formulation of the *distancing hypothesis,* and the introduction of the terms *distal* and *distancing,* are not only the product of the creative mind of an observant scientist, but also reflect the growing and inexorable interaction between understanding the origins of intelligent behavior, which we know as genetic epistemology, and the education of the child. The conclusion that Sigel has made a substantive contribution to the body of theory and research of genetic epistemology is inescapable, for he has brought about a recognition of the behaviors which indicate that the human learner has achieved representational competence, by which Sigel means that the child has learned to differentiate between reality and some representation of reality.

This contribution seems particularly meaningful in the area of early childhood education, which is recognized increasingly as at least in part the responsibility of developmental psychology. The use of representations of reality is a major teaching methodology employed in schools throughout the United States and other nations, often occupying a large percentage of the teaching strategies used in the social effort to insure the survival of neophytes; for example, the use of charts, pictures, films, and slides are all modes of representation. Certain judgments should be made about them in view of Sigel's observations as to their value, especially regarding a particular strategy as it reflects the reality of the child's experiential background.

The reality of the need to educate children, and the critical social need to intervene with compensatory education for some special populations of children, is a fact of life throughout the world. Sigel seems to have recognized this fact, which unfortunately does not impress some psychologists (Elkind, 1969) who seem to be willing to take a more maturational or *laissez faire* attitude toward early childhood intervention, a position which seems open to the charge of social irresponsibility, in view of current social circumstances and conditions. Fortunately, this position is not generally held in psychology, especially by those scientists attempting to develop research regarding such notions as distal stimuli, a concept which seems especially potent as psychologists attempt to come to terms with the sequential nature of the development of intelligent behavior, both within and outside the school.

8

∾∾∾∾∾∾∾∾∾∾∾∾∾

J. McVicker Hunt

∾∾∾∾∾∾∾∾∾∾∾∾∾

Few psychologists have so singularly concerned themselves with the development of intelligent behavior, and as a result so greatly influenced the emerging psychological frame of reference in this area, as has J. McVicker Hunt of the University of Illinois. Hunt's pervading concern, shared today by most psychologists, is with the *uniqueness* of human beings, and the singularity of the development of intellectual activity. His persistent theme is the influential role of *experience* on the development of intelligent behavior. Through his emphasis on experience, Hunt differs somewhat from Piaget, who identifies equilibration as the basic factor of developing intelligent behavior (Ripple and Rockcastle, 1964). Equilibration, according to Piaget, results as the organism adapts to the environment through accommodation and assimilation. Hunt's esteem for Piaget becomes obvious from an examination of his extensive and in-depth treatment of Piagetian theory (1961).

A study of Hunt's writings produces an appreciation for the individuality of man, sometimes missing in the work of earlier theorists, such as G. Stanley Hall's student Arnold Gesell (1945).

Adapted from "The Development of Intelligent Behavior VI: J. McVicker Hunt," *Psychology in the Schools,* VI, No. 2 (1969), 123–32.

Considering the genetic realities of human conception (that is, that in a single ejaculation of seminal fluid there may be as many as 200,000,000 sperm any one of which can fertilize the ovum, that at the moment of fertilization there are almost 17,000,-000 possible chromosomatic combinations, and that the merging of the maternal and paternal germ cells is a matter of chance), it seems impossible that individual human beings could be considered to be anything less than unique. Such was the case, however, and two basic misconceptions, predetermined development and the inherited nature and constancy of intelligence, tended to dominate much of psychological and educational thinking well into the twentieth century. These issues are still not put aside completely in spite of such sound refutations as that which appears in the early chapters of Hunt's scholarly text, *Intelligence and Experience* (1961).

Motivation Reinterpreted

In his article, "Experience and the Development of Motivation: Some Reinterpretations" (1960), Hunt reflected his interest in the changing beliefs about fixed intelligence and predetermined development, and clearly demonstrated his movement away from the S-R, drive-reduction position. Hunt does not limit his questioning attitude solely to the general concept of a drive-reduction model of motivation; he cites specific evidence calling into question the basic assumptions of such a model.

THE DOMINANT THEORY DEFINED

Tension reduction is the most basic assumption of both drive and psychoanalytic theories, which Hunt calls "the dominant." These models hold that every function of a living organism is to reduce or eliminate stimulation or excitation, and from this is drawn the concept that all behavior is motivated.

Hunt holds that the concept of all behavior as motivated implies that all living organisms would be essentially inert by choice, that is, whenever stimulated or excited the reactive organism would seek only tension reduction. This assumption is basic to the behavioristic notion of learned drives, and is fundamental to the psychoanalytic concept of anxiety.

Hunt questions that either interpretation is correct. To explain any behavior, the *behaviorist* is forced to rely on unobservable, internal or weak stimuli, whereas the theorist of a *psychoanalytic* orientation must necessarily assume some effort to escape from pain, and the *hedonist* seeks an explanation involving physiological or psychological imbalance.

A concept most often used in laboratory experiments and referred to by psychoanalysts in the realm of motivation is that of "learned fear" or "anxiety." Hunt does not question the realistic basis for (conditioned) learned fear drive or anxiety; in fact he cites his own work (1941, 1946) in this area. He does question seriously that conditioning as conceived by behaviorists (Dollard and Miller, 1950; Miller and Dollard, 1941) or anxiety as held by psychoanalytic theorists (Freud, 1936) are correct and complete explanations of an organism's developing fear of essentially indifferent or innocuous stimuli.

Behavior preferences or patterns are explained by S-R theorists as *habits*, whereas psychoanalytic theory emphasizes "modes." Habits are behavioral patterns which have served to reduce tension in the past in essentially similar situations. Hunt defines habit to include behavior fixated during infancy, ego-defenses, anxiety equivalents, cathexes, instrumental responses, and traits.

Change in behavior patterns is explained in these theories as motivation by punishment or homeostatic need, with reward given to the desired behavioral modification. Hunt holds that in both behaviorist and psychoanalytic theories, motivation is conceived as necessity, and "motivation means changing the emotional or drive conditions which are quite extrinsic to either the instrumental behavior or the cognitive, informational processes concerned" (1960).

Before stating his reinterpretations, Hunt acknowledges the massive experimental evidence to support homeostasis as well as conditioned drive and fear. He explains that his theoretical task is one of *reinterpretation*, rather than one of reformation or revolution.

HUNT'S REINTERPRETATION

Hunt first questions the validity of the assumption that all behavior is motivated and that without stimulation the organism is inactive or inert. Such a position attacks also the all-inclusive explanatory power of the homeostatic concept. To support his questioning attitude, Hunt refers to the work of Beach (1945), Harlow (1950) and Piaget (1952a), all of whom have reported observations of young animals or children engaging in activity during periods of low drive. Hunt brings out the point that strong stimulation during such a period tends to disrupt the low-drive behavior, and the organism moves to reduce the stronger drive.

Hunt objects to the "naming tendency" of some psychologists. He refers specifically to Berlyne (1954b), who has called this same phenomenon a curiosity drive. Hunt fails to mention White's (1959) competence drive, which seems to be little more than another attempt at naming-behavior. Hunt holds that simply naming the observable phenomenon explains nothing. We should, rather, admit that our observations contradict the dominant theories. Such an admission will leave the scientist free to embrace the concept of man as an *open energy system* in constant interaction with his environment, whose activity is modulated but not initiated by the stimulus situation. Hunt's entire reinterpretation rests on his belief that the activities of man do not require homeostatic need nor painful stimuli, but rather that *to be alive is to be active.*

This concept of life and activity necessitates the rethinking of current notions of reinforcement and the theory that conditioned fear and anxiety are always resultant from painful stimulation or homeostatic need. Hunt maintains that fear and anxiety have some other basis than pain or need, and turns to the *incongruity* or *dissonance theory* which he attributes to Hebb (1946, 1949).

THE INCONGRUITY-DISSONANCE PRINCIPLE

A portion of the experimental evidence for the incongruity-dissonance principle comes as the result of observing *withdrawal behavior* (fear) in young chimpanzees when presented with some familiar object or person changed in some important, unexpected way. For example, the young chimps reacted with fear (withdrawal behavior) to their keeper wearing a Halloween mask. Hunt holds that the withdrawal response may be unlearned in the sense that it may be a reaction due to incongruity of the presented object with the learned aspects of the familiar object. Familiarity is then understood to be the result of experience. In Hebbian terms, a central process has been established by frequent experience with the redundant aspects of an object. Presentation of the familiar object in unfamiliar guise results in new receptor inputs which are incongruous; therefore, the organism fears the object. Hunt concludes that until the central process has been developed (learned) incongruous stimulation is impossible, and without incongruity fear does not occur.

The incongruity-dissonance principle is applicable to much research regarding the fear response, and sheds new light on work such as Lewin's positive and negative valence (1935), and Miller's conceptualization of approach-avoidance gradients (1951). In each instance, *after* an object or place has become familiar to the organism, incongruous or dissonant stimuli are introduced resulting in withdrawal behavior. It is interesting to note that the slopes of Miller's approach-avoidance gradients are a function of the degree of incongruity or dissonance introduced into the environment, and if incongruity is introduced in sufficient degree, avoidance behavior tends to disrupt approach.

A similar interpretation of Freudian fixation is possible. Fixation is generally conceived to be either an arrest of libido at an immature stage of development or its strong attachment to early and inappropriate objects. Learning theorists Dollard and Miller (1950), Sears (1951) and Whiting and Child (1953) have elaborated this psychoanalytic concept in behaviorist terms, as well as considering the defense mechanisms which are postulated to reduce anxiety. In such instance, fear results in fixation, and comes about as the result of the introduction of dissonant incongruous stimuli in an otherwise familiar environment. For example, a child may learn to fear a puppy when the animal charges excitedly into the child's environment, barks, and leaps upon him, sending the child sprawling on the floor. The child can learn to enjoy a puppy, however, if their introduction is calm and supportive, without excessive exuberance. Then the two of them can learn to make dissonance together in the house.

Hunt holds that the incongruity-dissonance principle explains Hebb's theory of cell assemblies (1949) and Piaget's notions of adaptive behavior with the invariant functions of accommodation and assimilation (1952, 1954).

He also applies the principle to Helson's model of adaptation level (1947, 1948), Festinger's cognitive dissonance (1957), Rogers' discrepancies in the phenomenological field (1951), and Kelly's system of personal constructs (1955).

Hunt believes that the incongruity-dissonance principle makes motivation and reinforcement intrinsic to the organism-environment interaction. He compares this to the information-processing model, in which a feedback system provides the organism with the knowledge of dissonance and its source. The organism then acts to resolve dissonance.

Intelligence and Experience

Hunt's most definite statement concerning the development of intelligent behavior was published in 1961, and represents a monumental effort to resolve the lingering effects of the misconceptions of fixed intelligence and predetermined development. Hunt explores the recent transformation of psychological theorizing and investigation concerning the development of intelligent behavior and its relation to experience, and holds that the emerging evidence must bring about a recognition of the *central processes* of intelligence, and the *impact of experience* on the development of central processes.

Hunt in his preface calls his text "a kind of case history in behavioral science, presented before all of the case has become history" (p. v). Nonetheless, the depth of this important "case history" is revealed in Hunt's careful analysis of the belief in fixed intelligence and predetermined development (pp. 10–64). His refutations of these misconceptions are thoughtful and completely defensible from both an experimental *and* a logical frame of reference.

Hunt is one of the few major theorists in psychology who not only calls on the massive support of experimental evidence to support his conceptions, but proceeding according to the scientific method, also uses logic to establish the rational supports for his postulates. Piaget is another such theorist. Unlike Hull, who was also able to use both edges of the scientific sword, Hunt and Piaget more clearly manifest their concern for human development. This willingness to accept support from all acceptable sources seems to define most clearly a critical component of the emerging *zeitgeist* of educational psychology.

After a very thorough examination of information processing and experience, and Piaget's theoretical developments, Hunt exposes his own thoughts in the chapter, "Some Reinterpretations" (pp. 308–46).

INTELLIGENCE TESTS

Hunt holds that intelligence tests consist essentially of samplings of behavior, and that traditional intelligence tests typically sample skill categories, verbal and mathematical in particular. The efforts of Spearman (1927), Thurstone (1938), and Guilford (1956) have yielded "systems of coordinates" (p. 311), which Hunt believes have little or nothing to do with the natural structures of the development of intelligent behavior.

Hunt states that the DQ and IQ are fluctuating rates relative to age, most useful for the comparison of individuals (1961, p. 312). Hunt postulates this concept of a *fluctuating* rate in opposition to the concept of a *constant* rate which would be an expression of the notion of fixed intelligence. Hunt mildly refers to the fluctuation of DQ and IQ as "embarrassing" to older theories of intelligence and development. To the authors, the evidence seems more crippling than embarrassing.

Hunt explains his opposition to fixed intelligence and predetermined development by discussing the unfortunate consequences of such beliefs (pp. 345–46). The first detrimental consequence is to encourage the practice of leaving infants essentially unstimulated during the earliest months of life. In an equally devious way, these misconceptions discourage investigating the effects of various programs of child-environment interaction during the course of development.

The belief that early experiential impoverishment appears to inhibit the development of intelligent behavior has been instrumental in shaping Hunt's positions, and is reflected in the emerging developmental theory. Experiential deprivation has produced deviant maternal behavior (Birch, 1956) and ineffective nest-building behavior (Reiss, 1950) in rats, disoriented feeding behavior in chicks (Cruze, 1935), and seriously delayed development in institutionalized children (Dennis and Najarian, 1957). Rosenzweig (1966) has published evidence of increased cortical mass as the result of experience, and the evidence continues to support the transformation which so completely occupies Hunt's attention.

The Psychological Basis for Using Pre-school Enrichment as an Antidote for Cultural Deprivation

In this important article originally published in the *Merrill-Palmer Quarterly* (1964), Hunt turns most forcefully to the crucial role of experience in the development of intelligent behavior. Hunt believes that there is evidence of a radical change in the current conceptions of human development, and classifies six specific areas of innovation. The first two changed beliefs are the concepts of fixed intelligence and predetermined development, explored in depth in *Intelligence and Experience* (1961). Third, the belief in the fixed and static, telephone-switchboard nature of brain function has changed. The theories that experience is unimportant during the early years and that whatever experience does affect later development is a matter of emotional reactions based on the fate of instinctual needs are rarely held today among professionals. Finally, important rethinking has been done regarding the belief that learning must be motivated by homeostatic need, by painful stimulation, or by acquired drives based on them. This final consideration was the subject of Hunt's reinterpretation of motivation (1960).

THE STATIC SWITCHBOARD CONCEPT

The concept of the brain function as static originated when Darwin shifted his concern from physical to mental evolution (1872), a development which provided the stimulus for comparative psychology, the purpose of which was to demonstrate a relationship between the mind of man and that of lower animals. Romanes (1883, 1884) accepted the concept and generated it to anthropomorphism. The concept was interpreted in this light until it was scientifically purified by C. Lloyd Morgan's (1894, 1909) deft surgical use of Occam's Razor.

Attacks by Thorndike and Woodworth (1901), reduced the influence of faculty psychology, until Hull (1943) elaborated on the process of trial-and-error learning, attributing to the organism a set of innate response tendencies. S-R chains were formed in order to explain complex human behavior. The telephone was a convenient and timely model, and its switchboard became a favorite analogy, so that the brain was conceived to function as a connector, filling an otherwise empty cavity.

To account for symbolic behavior, Hull (1931) proposed the r_g–s_g mechanism to be a pure stimulus act when it served to direct behavior. Miller and Dollard conceived response-produced cues (1941), which Osgood interpreted as central mediating processes (1952).

Hunt holds that the theoretical need for active brain processes was shaped and stimulated by Wiener's cybernetic model

(1948). Additional developments have come about because of the work of the information process theorists such as Newell, Shaw, and Simon, who specify three requirements for problem solving: *memories, operations,* and *hierarchical arrangements* of the memories and operations (1958). Hebb's observations based on findings in neuropsychology (1949), appear significant to Hunt, who conceives the function of early experience to be one of *programming the intrinsic portions of the cerebrum for future functioning in learning and problem solving.*

THE ROLE OF EARLY EXPERIENCE

One early misconception attributed little importance to early preverbal experience, for it was believed that early experience had little or no effect on adult behavior. Freud was little heeded in his emphasis on childhood experience, especially "synchronized experiences" from the environment (Hall, 1954, p. 97). Freud held these experiences to be an important factor in the development of a healthy ego.

Hunt states that there appear to be two effects of early experience. The first is *the averting of painful stimulation,* and the second is an *increase in the capacity to learn.* This second effect of experience becomes increasingly important in light of the need of very young children for a wide variety of looking and listening experiences. Hunt finds additional support for this need for experience when he refers to Piaget's (1936) statement that the more a child experiences visually and auditorially, the more he will tend to seek visual and auditory stimulation.

AN ANTIDOTE FOR CULTURAL DEPRIVATION

Hunt calls his ideas a "theoretical fabric" which can be applied to pre-school enrichment programs for the culturally deprived. Cultural deprivation is defined as "a failure to provide an opportunity for infants and young children to have the experiences required for adequate development of those semiautonomous central processes demanded for acquiring skill in the use of linguistic and mathematical symbols and for the analysis of causal relationships" (Hunt, 1964).

Hunt believes that the early work of H. M. Skeels and his

associates is relevant and worthy of reinterpretation (see Skeels et al., 1938, 1939; Wellman et al., 1940). He finds Oscar Lewis' anthropological studies of the culture of poverty (1961, 1966) also important and meaningful. Hunt holds that one of the most significant factors in lower-class life is crowding, which though it may facilitate development in the first year, becomes inhibitory thereafter. The environment of poverty is often noise-filled but nonverbal, and there are few models for learning vocal language. Often in the culture of poverty, the child's inquiries about the world which surrounds him are negatively rewarded. There are few toys in his environment. Most of the people, especially the adults with whom he will interact, are fixed on the monumental task of survival. For these reasons, Hunt envisions the onset of retardation after the first year and from then on its effects are cumulative. Hunt believes that retardation due to deprivation can probably be reversed to a marked degree. He voices support for early childhood education, expressing the opinion that it would be better to begin with children in their *third* year.

Basing his proposals on the work of Piaget (1936), Hebb (1949), and Hebb and Thompson (1954), Hunt suggests that the most important factor in a pre-school program as an antidote for cultural deprivation is the opportunity to encounter circumstances which will foster the development of the central processes of intelligent behavior. These encounters will tend to facilitate the development of imagery, which is ordinarily followed by the development of the language skills so important in formal education.

The children of poverty need a wide, varied range of encounters which will compensate for the deficiencies of their cultural environment. These experiences should be within the child's ability, or synchronized, in Freudian terms. Hunt suggests that psychologists and educators might for these reasons be wise to look again at the methods developed by Maria Montessori, who based her teaching methods on the spontaneous interest of children in learning.

Montessori stressed observing children to foster their interests, and the training of what we understand as information processes. Hunt states that the chief advantage of Montessori's techniques was that she afforded each child the opportunity to discover the circumstances to match his interests and stage of development.

Some Reinterpretations of Traditional Personality Theory

Throughout his theoretical fabric weavings, Hunt has been possessed by an overriding concern for reality. In some way this concern may account for his role in shaping the current psychological viewpoint regarding the development of intelligent behavior. His focus on the role of experience becomes more logical due to this reality-orientation. Hunt has little patience with the misconceptions of fixed intelligence and predetermined development, and his concern has led to his being criticized as concentrating on these misconceptions *ad nauseam* (1965, p. 84). The authors disagree with this criticism, believing that the long-term effects of these misconceptions still contaminate certain areas of psychological thought.

Hunt in his *American Scientist* article (1965) reinterprets six beliefs of many personologists. He first holds that traits alone are not the major source of behavioral variance. Rather, emerging evidence indicates that it is neither "individual difference, *per se,* nor the variations among situations, *per se* that produce the variations in behavior. It is, rather, the interactions among these which are important" (p. 83).

In order to understand the variations in the meanings of situations to people and the variations in the modes of response they manifest, Hunt suggests we look toward instruments such as Osgood's Semantic Differential, which classify people in terms of the kinds of response they make in various categories of situations.

MOTIVATION OF BEHAVIOR

Hunt rejects the traditional models of motivation, e.g., drive theory, hedonism, and homeostasis, as incomplete explanations. Hunt postulates that new evidence combines to indicate that a *system and mechanism of motivation is inherent in the organism-environment interaction*. The determination of the essential characteristics of this mechanism of motivation is very difficult, but Hunt identifies *incongruity*, by which he means the discrepancy between informational input and stored cognitive data. He refers (1965, p. 85) to personal communication with Isaacs, who has made a further distinction with which Hunt appears to agree. Isaacs holds that *novelty* is the discrepancy between mere informational input and stored cognitive data, while *incongruity* is the more serious discrepancy between input and the individual's established commitments and plans.

Regardless of the essential character of interaction, Hunt holds that there is an optimum which is a function of experience, and relates this to Helson's adaptation model (1959). He also indicates his belief that defense mechanisms may function to protect the individual from extremely incongruous informational input.

THE ROLE OF COGNITIVE FACTORS IN DEVELOPMENT

Hunt attributes to Freud the major influence for the recognition of the role of early experiences in development (1965, p. 87). However, a maximum of importance was given to emotional factors to the neglect of cognitive factors during the preverbal years. This misconception has given support to the theories of predetermined development and fixed intelligence, whereas emerging evidence indicates that just the opposite should hold.

Evidence supporting the need for a wide variety of experiences in the early years comes from animals, and the higher the animal on the phylogenetic scale the more clear cut is the evidence. The importance of early cognitive experience "increases as that portion of the brain without direct connection to sensory input or motor outlet increases relative to the portion which does have direct sensory and/or motor connections (i.e., with the size of what Hebb (1949) has termed the A/S ration)" (1965, p. 88).

In turn, Hunt considers whether emotional attachments derive from gratification of libidinal or homeostatic needs. He proposes, in line with increasingly dominant central process theories, that organisms become attached to objects which have become recognizably familiar in the course of *repeated encounters*, and show distress when the attached object is removed from the perceptual field. Interestingly, this appears to occur at about the time which Piaget (1936) believes the human organism develops the imagery necessary for future intelligent behavior. Hunt also holds that the ethological phenomena of imprinting (Hess, 1962) may be a special case of emotional attachment deriving from recognitive familiarity.

EARLY PAINFUL STIMULATION AND ANXIETY

Hunt rejects the "trauma theory of anxiety." This theory assumes the conditioning concept of fear by suggesting that painful stimulation or strong homeostatic need leads to a proneness to sensitivity and anxiousness. Though this concept is inadequate in view of much emerging evidence, it is still widely held, especially among clinicians of a psychoanalytic orientation. Hunt holds that the evidence at least lessens the explanatory power of the trauma theory, and that painful stimulation may not necessarily condition fear. Rather, he believes that painful stimulation may serve to raise the adaptation level and in so doing reduce its aversive nature.

He does not commit the error of the trauma theorists by maintaining that his interpretation is the *only* source of anxiety, attributing a residual role to the trauma theory. He calls, rather, for a clearer definition of the types of experiences which produce anxiety in young children.

Hunt began his treatment of traditional personality theory by elaborating on the process of science as consisting of the "creating" of conceptual schemes in which the relative validity of competing concepts is tested and new concepts are formulated or emerge. The effects of the scientific process have been slow to touch the science of personology, but as he summarizes Hunt seems to capture the essence of the *zeitgeist* of psychology that perhaps ". . . the yeastful and self-corrective dynamic of science has at last found its way into knowledge of persons and of personality development."

9

~~~~~~~~~~~~~~~~~~~~~~~~

The Central Process
Theorists

~~~~~~~~~~~~~~~~~~~~~~~~

The 1940's ushered in a period of significant breakthrough for the psychologist and educator concerned with the development of intelligence. This era was one of radical change and departure from previously held conceptions, particularly from the notions of fixed intelligence and predetermined development. There had long been evidence which did not support the time-dependent, hereditarian views of G. Stanley Hall, his student Arnold Gesell, and others. The dissonant evidence from studies of separated identical twins, the inconstancy of the IQ, and the effects of education were rationalized away by these theorists who exploited a normative approach to human development.

Disagreement with the concept of fixed intelligence was not new; it was simply ignored. In fact, as early as 1909, Alfred Binet protested that the theory was "brutal pessimism." The champions of the concept, however, were respected and powerful men such as Francis Galton, Cyril Burt, and James McK. Cattell, who found the roots of their creed in Darwin's theory of natural selection. Fixed intelligence represented a momentary zenith for the nature side of the devious and complicated nature-nurture controversy.

Adapted from "The Development of Intelligent Behavior V: Central Process Theorists," *Psychology in the Schools*, V (1969), 24–37.

Today's emerging views are rooted in psychologists' attempts to cope with the realities of the human organism. Man is not a machine, nor a telephone exchange center. He is, rather, a problem-solver, an organism which considers alternatives and formulates hypotheses, as postulated by Edward C. Tolman and Egon Brunswik (1935) and I. Krechevsky (1932).

Many behaviorists such as Irving Maltzman (1955) and Tracy Kendler (1963) support a concept of a mediational process, following Clark Hull's theoretical statements regarding pure stimulus acts (1931). Later behaviorists, Neal E. Miller and John Dollard (1941), postulated response-produced cues in another attempt to add dimension to understanding the development of intelligent behavior. A further behaviorist-oriented development came about when C. E. Osgood attempted to bridge the gap between cognitive and behaviorist theories with his unique theory of mediational processes (1952). Osgood's work seems to have influenced and to have been extended by cyberneticists (Newell, Shaw, and Simon, 1958) and information-process theorists (Miller, Galanter, and Pribram, 1960), who are currently prominent in studies of computer-simulated intelligence.

Harry F. Harlow

Previously acquired learning, which has been accommodated to and assimilated into the cognitive structures of the organism, determines the usefulness of all new learning. Adaptation to the learning situation and the achievement of facilitative skills in learning are called *learning sets*. A student learns how to think through a problem, or to identify the statistical technique which can best be used to exploit data. A busy professional learns to identify and extract the essential portions of a journal article. All of these commonplace learning "shortcuts" and methods of attack are exemplars of learning to learn, i.e., of responding to relevant cues in the learning process.

Learning to learn was not a new phenomenon at the midpoint of the twentieth century, but it was commonly ignored. Notable exceptions to this trend were the studies of twins by Newman, Freeman, and Holzinger (1937), the Berkeley Growth Study (Bayley, 1949) and the Berkeley Guidance Study (Honzig, MacFarlane, and Allen, 1948) in which wide variations in intra-individual IQ were reported from longitudinal research.

Nevertheless, resistance was potent and exacting, as can be seen in McNemar's (1940) reanalysis of the data from the orphanage studies of Skeels et al. (1938), which reduced their conclusions to nonsignificance.

The emerging *zeitgeist* gathered momentum, and a major breakthrough came in 1949, when Harlow at the University of Wisconsin reported initial evidence that an organism (in his research Harlow used rhesus monkeys) *learned how to learn.* After repeated experiences with a given kind of problem, the organism developed a "learning set." In Harlow's experiment, the monkey was put into a forced-choice situation, in which he had to make a decision in favor of one of two objects. The objects were marked with different colored covers on cups. Although one *color* was consistently rewarded, the *position* of one object in relation to the other object was random but predetermined. In the first phase of the learning experience, the subject tended to respond to *where* he found the last reward (a grape), but learned gradually to ignore the place cue and respond to the color cue. The animal had to discriminate between relevant and irrelevant dimensions of the problem, to learn to respond to the relevant stimulus, and to inhibit a tendency to respond to the irrelevant stimulus (the place of last reward).

Following the first experiments, Harlow continued his studies, presenting a set of eight monkeys with a series of discrimination problems. The monkeys were given 50 trials to learn the discrimination of one color from another, then another 50 trials to learn to discriminate between two different colors. Another problem involved discriminating a particular container by shape, i.e., square versus round. The results confirmed Harlow's earlier hypothesis that an organism develops "learning sets," i.e., that one can learn to learn. In the first three trials, the organism's percentage of successful discriminations was not significantly greater than that attributable to chance. Successful discriminations were made in only about 75 per cent of the instances after six trials, but the number of successful discriminations *after experience with the problem* rose dramatically. In only the first instance was the percentage of correct discriminations equivalent to or less than that explainable by chance. In following trials there was a sharp increase to almost perfect discrimination. The animals learned how to solve discrimination problems. Harlow's "educated" monkeys eventually needed only one trial to make the proper solution.

Using the same set of monkeys, Harlow complicated the environment and demonstrated that organisms can *learn to solve reversal problems* (1949). The reversal problem involved switching the relevant stimulus after an initial learning experience; in the first few trials the color of the cover was rewarded, but a reversal was made and the unrewarded color became the relevant and rewarded stimulus. At the point of reversal, the monkeys made many mistakes in discrimination, but after experience with the problem, the percentage of correct responses rose to 97 per cent correct. In addition, having learned the reversal problem, the monkeys were eventually able to achieve almost perfect performances in similar problem situations after the first trial.

Further extending his work, Harlow applied his concept of *oddity problems* (1951). In this kind of problem, the organism was presented with three stimulus objects, two of which were identical. The organism had to ignore previous relevant stimuli of perceptual cues (color) and position, and learn to respond to one shape in comparison to two other shapes which themselves had to be compared. As was expected, the monkeys made many errors in the initial phases of the problem, but they showed progressive transfer from their previous experience with discrimination problems to the situation in which they were placed. At the end of Harlow's training experiences, the monkeys were making 90 per cent correct discriminations on the first trial.

That learning sets may operate as relatively *isolated central units* was demonstrated by Harlow in an experiment with six monkeys who had had previous experience with discrimination training, but no experience with positional discrimination (1949). After an initial block of problems involving object discriminations, the monkeys were presented with an equal block of problems involving right-left positional discriminations. Twenty-five blocks of fourteen problems alternated between object discrimination sets and positional discrimination sets. The results showed that the original object discrimination remained at a high level of accuracy after a small decline with the second set, results which may be explained as interference from the first set of positional discrimination problems. Throughout the experiment, the performance on the previously unexperienced positional discrimination problems improved steadily until on the last set of such problems the animals were making 85 per cent correct discriminations.

Harlow (1951) commented on the relative ease and efficiency with which the animals were able to move from one type of problem-solving situation to another, proposing that learning sets are functionally isolated and are a basic mechanism in problem solving in complex situations. He remarks that man, to a much greater degree than rhesus monkeys, has the ability to select alternative solutions to problems in relation to the situation confronting him.

Harlow and Harlow summarized their research writing:

> Thus the individual learns to cope with more and more difficult problems. At the highest stage in this progression, the intelligent human adult selects from innumerable, previously acquired learning sets the raw material for thinking. . . . Thinking does not develop spontaneously as an expression of innate abilities; it is the end result of a long learning process. . . . An untrained brain is sufficient for trial-and-error, fumbling-through behavior, but only training enables an individual to think in terms of ideas and concepts [1949, pp. 38–39].

Man learns to think and to solve the problems encountered naturally, and those which are put into his environment by calculated intervention, such as the educational encounter. His learning can be facilitated, if he is given the opportunity to practice manipulating his environment in both physical and intellectual situations, so that he may more efficiently develop learning sets.

Hebb's Neuropsychological Theory

The early conceptualizations of Donald O. Hebb of McGill University were dominated by a scientific desire to understand and explicate one domain of behavior: human motivation. The gradual extension of Hebb's ideas to the development of intelligent behavior provides a model of change in the science of psychology based on a specific criterion. In Hebb's case, the changes came as a result of a conscientious attempt to correlate psychological theory with new concepts in neurology and physiology.

In an address, "Drives and the C.N.S. (Conceptual Nervous System)" (1955), Hebb explained that his notions of motivation had emerged progressively as a result of experimental evidence, and in response to more recent conceptions of the nervous system. Hebb's theory contains many hypothetical constructs, such as the *cell assembly* and *phase sequence,* and has generated much research in both physiology and psychology.

Hebb considers himself a behaviorist; however, his theorizing is admittedly interpretative. Thus, Hebb might be accused by other behaviorists such as B. F. Skinner (1950) of seducing psychology into forgetting its primary task of accounting for behavior and producing "wasteful" research. Hebb appears to have been little influenced by such criticisms.

THEORETICAL EVOLUTION

The first considerations of motivation by behaviorists followed the lead of the natural sciences, especially physics and chemistry, in seeking the cause of action *outside* the organism. The human being was regarded as passive until acted upon by external forces—sources of power which were called *drives.* There were the primary drives: sex, pain-avoidance, hunger, and thirst; and learned or derived drives, sometimes called secondary drives. Within this concept of human motivation, the nervous system was seen as inert unless activated from the outside. This concept is still prevalent, in spite of recent developments.

Hebb saw many difficulties with drive theory, especially since it did not adequately account for the fact that organisms were active even when not under the pressure of drives. Men learn while exploring their environment, and eventually psychology was forced to come to terms with the realities of human behavior. Hebb was one of the leaders in this movement.

Proposals in his book, *Organization of Behavior* (1949), were based on a more recently developed conception of the central nervous system. The nerve cell was no longer believed to be physiologically inert, responding only to external stimulation. Rather, it was understood to be alive and active. For Hebb the problem was not to account for the energizing of behavior, but rather to explain its *patterning* and *direction.* Hebb hypothesized that brain activity determined behavior, and that the problem was to account for *inactivity.* This new posture accounted for a great deal of information about human behavior, but experimental evidence soon convinced Hebb that the conception needed further revision.

In Bexton, Heron and Scott's experiment (1954), students were paid to remain in bed with as little stimulation as possible. They were allowed time only for toilet and food. The subjects wore goggles, gloves, and cardboard cuffs, and to the greatest degree possible auditory stimulations were controlled. The students were paid $20 a day, and were encouraged to remain with the experiment as long as they wanted. Few chose to remain more than a few days.

This classic experiment was extremely debilitating to Hebb's concept of motivation at that time, and was pivotal in his revision efforts. He reasoned that if the thought processes and motivation were internally organized (as he had conceived them to be until that time), there would be no reason for them to break down because of sensory deprivation, as the experiment so clearly demonstrated they did.

The students involved in the experiment had become restless after only eight hours of deprivation. They developed a need for stimulation, manifested signs of behavior disorganization, and developed symptoms of psychological impairment. For Hebb, the experiment was a complete contradiction of his theory. To find an answer, Hebb turned to recent findings in brain physiology. Of unique significance was the discovery that the brain stem functions as an *arousal system*. The reticular formation of the brain stem had been known for some time, but its function had remained a mystery.

An influential work regarding the reticular formation was the discovery by Moruzzi and Magoun (1949) that electrical stimulation of the reticular formation produced an activation pattern in the EEG. Lindsley, Bowden and Magoun (1949) worked with cats' brains to produce opposite effects.

Activation of the reticular formation was seen to have two major sources: *sensory stimulation* and *cortical impulses* into the reticular formation. This second source of stimulation was demonstrated in psychological and related research and theory (see French, 1957; French et al., 1955; Hebb, 1964; Hyden, 1967; Lindsley, 1950, 1951, 1957; and Lindsley and Bowden, 1949) in which stimulated cortical areas were noted to have increased activity in the reticular formation.

Earlier conceptions held that sensory input from the receptors, such as the eyes and ears, went relatively directly to the sensory areas of the cortex via specific pathways. The new conception postulated two pathways for different (periphery to central) stimulation, the first of which coincides with the early notion of direct sensory pathways. The second pathway for incoming sensory input is via the arousal system in the brain stem, which results in non-specific stimulation of the cortex.[1] This second pathway does not provide information, but rather serves to tone up and arouse the cortex and to alert it for action. In experiments in which the second pathway was surgically blocked, the animal became inert and entered a coma.

Hebb's continuing concern with the relationship of cerebral function to behavior was expressed earlier with regard to cortical damage and the retention of learning and acquisition of new abilities (1942). Hebb held that although only part of the brain may be required in order for one to *maintain* an ability once it has been learned, the initial learning proceeds much less efficiently if the brain is not intact. For example, loss of tissue might have no effect on a learned skill in an adult, whereas a similar loss of tissue in a child would interfere significantly with his acquiring that same skill. Hebb also proposed, however, that damage to the speech area of the cortex in an adult might result in aphasia, whereas similar damage to a child would have less effect on the acquisition of speech, since other areas of the brain would tend to compensate by taking over the normal functions of the damaged area.

Hebb suggests two results of a sensory event. The first is a guiding function for behavior which Hebb calls a *cue function,* and the second is the *arousal function,* less noticeable, but no less important in its role of arousing and alerting the cerebral cortex.

With the development of this theory, Hebb has returned to drive theory, conceiving drive as the energizer of behavior. This is, however, a much more sophisticated concept than the earlier drive (tension reduction) model. Hebb proposes that with this new conception, drive theory can account for such things as the positive attraction of risk-taking and thrill-seeking behavior which seem to fascinate human beings.

CENTRAL PROCESSES

In *The Organization of Behavior* (1949) Hebb recognized that phenomena labeled *set, attention, attitude, expectancy, hy-*

[1] Ascending and descending projective fibers link the cerebral cortex with the RAS or reticular activation system (related to sleep, arousal, and attention) which provided a referent for Lindsley's "activation theory of emotion" (in Cofer and Appley, pp. 398–407). Thus the RAS and limbic systems, both associated with the hypothalamus, mediate emotional experience and expression.

pothesis, intention, vector, need, and the like, have one and only one common element. The crucial influence is not an immediately preceding sensory stimulation for each response, but rather, an ongoing central activity which Hebb labels the *autonomous central process* (p. 5). Hebb concludes,

> The problem for psychology then is to find conceptions for dealing with such complexities of central neuron action; conceptions that will be valid physiologically and at the same time "molar" enough to be useful in the analysis of behavior [p. 11].

One of the basic assumptions made by Hebb in his theoretical text (1949) has to do with growth processes occurring at the synapse between two neurons. From this assumption, Hebb draws specific postulates concerning the anatomical organization of the nervous system.

Reverberatory loops. The nervous system is held to be a collection of individual cells, arranged so that there exist loops of several cells through which excitation can reverberate, although these loops do not remain active indefinitely without additional excitation. The loops, which allow continuing activity long after the cessation of the original stimulation, are called *reverberatory traces* or *holding mechanisms* (1958). Hebb believes that the reverberatory trace cooperates in the structural changes necessary for permanent or long-term memory, by temporarily carrying the memory until the structural growth changes can be made, then releasing the short-term memory to the more permanent structure. Hebb calls this the *dual trace mechanism* (1949, p. 61).

The growth principle. Hebb assumes that the reverberatory trace induces cellular change adding to its stability. He says:

> When an axon of cell A is near enough to excite a cell B and repeatedly or persistently takes part in firing it, some growth process or metabolic change takes place in one or both cells such that A's efficiency, as one of the cells firing B, is increased [p. 62].

Such a process of development might be the result of physiological proximity between an axon of A and an axon of B, and also the efficiency of the chemical transmitter substance might become greater.

The cell assembly. Hebb proposes the cell assembly as the basic units of the functional organization of the brain. When a specific cortical area is excited, the preponderant effect is a fanning out of excitation to other regions. Patterns of excitation develop as a result of "association" and facilitate excitation of one another. A simple closed circuit (it may help if we visualize a three-dimensional unstructured lattice) develops and excitation occurs through a process of convergence, i.e., two excited cells converge on another cell. The converging cells need not have anatomical or physiological relation to one another. The loop of cells is postulated to include *motor cells,* thereby incorporating motor activities in perceptual learning. Repeated activation of a pattern binds a set of cells together so that eventually it may function as a unit. This functioning loop of cells constitutes the cell assembly, a basic building block in Hebb's theory.

The phase sequence. Perceptual learning occupies a position of importance in Hebb's theory. He assumes that most early perceptual learning involves the development of cell assemblies which represent lines and corners (in the learning of shapes) and that a special cell assembly is developed for each type of corner and line. Phase sequences are particular orders of cell assemblies *and* receptor involvement. For example, in order for a square to be a *complex perceptual entity,* a person will focus his receptor (probably his eyes, or sense of touch if he is blind) on corner A, and presumably cell assembly *a* will be activated in the cortex. The receptor moves to line AB, and in turn cell asssembly *ab* is activated, and so on until the shape is understood. Feedback from the receptors and receptor movement aids in the development of a neural counterpart of the perceived square. The sequence of the cell assemblies in perceiving the square is integrated into the phase sequence.

Repeated experience connects cell assemblies into more elaborate perceptual units, and synaptic growth occurs between neurons involved in the cell assemblies and the receptor movement. Eventually, activation of cell assembly *a* will activate the entire phase sequence for the square ABCD and the organism will *short circuit* the cell assembly chain. Short circuiting is evidence of more efficient learning.

By emphasizing that movements develop with perception and in accordance to the neuronal growth principle, Hebb has eliminated the need for a *homunculus* to explain response to stimulus. Consequently, the motor and cognitive aspects of perception may be understood as the neural basis for the phenomena Harlow identified as *learning set.*

The Multidimensional Nature
of Human Abilities

In a report for the Committee on the Identification of Talent formed by the Social Science Research Council in 1951, Mc-Clelland and his associates (1958) initially concentrated upon values and motives (the nonacademic determinants of achievement) as well as social skills and occupational status (the nonacademic types of achievement). For the report, however, Baldwin prepared a chapter upon the role of an "ability" construct in a theory of behavior (pp. 195–233). He started with the notion that any ability attributed to a person reflects what he *can* do. In the language of Baldwin's chapter, the behavioral capability is a precondition to running a four-minute mile. Nevertheless, whether or not a runner attains that performance in any given race depends upon other factors—cognitive (planning), situational (conditions), and motivational (incentives).

In general, cognition (knowledge-ordering behavior) integrates a multitude of sensory cues (sorting information from error) and provides an internal "schema" to guide motoric actions without having to rely upon external guiding cues. Accordingly, Baldwin (p. 231) concludes that an ability linked with cognitive guidance does not necessarily depend upon any single sensory cue for feedback; for example, compare a man's movements in his own dark bedroom (cognitively mapped) and in a strange, darkened hotel room. Contact with one object whose relation to others in one's own room already is known permits a person to "place" other objects in the familiar room. Guidance provided by external sensory feedback, in a new or changing situation, however, permits adaptive behavior under shifting conditions. Finally, Baldwin recognized three sources of correlation between abilities: identical behavioral components, intrapersonal patterning, and cultural expectations (again in the language of Baldwin's report).

FERGUSON'S THEORY OF HUMAN ABILITIES

Ferguson's conceptual framework (1954, 1956) links the study of human ability with the study of human learning and the concept of transfer. The main points of Ferguson's theory fit with and provide a reasonable explanation of many reported observations in school settings and experiences in research (McGuire, 1945, 1949, 1956).

1. *The abilities of man* are attributes of behavior which, through learning, have attained a crude stability or invariance not only in the adult, but also in childhood and adolescence. This statement holds true unless an individual encounters learning experiences calculated "to make a difference" in his or her behavioral capabilities. In the initial proposal the planned interventions were termed educational *teleses* (Hindsman and Duke, 1948).

2. *Biological factors* in the formation of abilities fix limiting conditions. Parenthetically, the notion of *three interacting environments* advanced by respected geneticists in the Messenger Lectures on the Evolution of Civilization at Cornell University (Muller, Little, and Snyder, 1947) provides a reasonable heuristic device to replace the heredity versus environment distinction. Three levels of environmental influences operate from conception until death to determine the development and individuality of human beings (pp. 89–108).

 a. The *morphogenetic* or gene-controlled environment is a compromise between the two gene systems present (female ovum, male sperm) at the conception of an individual human replacement.[2] Although behavioral scientists have been unable to devise means for the assessment of potential (see pp. 118–19, for a discussion of "convenient fictions"), the view is widely held that "the genes fix limits, perhaps rather broad, of potential intellectual development" (Ferguson, 1965, p. 40).

 b. The internal or *neuro-endocrine* environment is initiated when placental hormones take over some developmental control and regulatory processes earlier mediated by nucleus-cytoplasm and cell-cell interchanges. Morgan's third edition of his *Physiological Psychology* (1965) has clear accounts of neuronal

---

[2] Since Linus Pauling won a Nobel Prize in 1950 for work on the atomic structure of protein, research has shown that a gene is a segment of deoxyribonucleic acid (DNA) which carries the blueprint for making new proteins. Ribonucleic acid (RNA), occurring as nucleotides in the cytoplasm of cells as well as in the form of "messenger" and "transfer" RNA's, evidently provides a bridge between DNA master molecules and proteins formed during the life cycle of living organisms (Beadle and Beadle, 1966, pp. 180–91). Nevertheless, subsequent research does not substantially contradict the late Professor Muller's account of "the work of the genes" in the 1947 book (pp. 1–65), including the operation of nucleoprotein and nucleotides (which he regarded as essential agents for transfers of energy in living systems).

physiology and the internal environment (pp. 61–108) as well as emotional and other behavior (sleep, arousal, activity) associated with a reticular activating system (RAS) (pp. 41, 327–48) which may be linked to motives as "affectively toned associative networks" (McClelland, 1965). Psychological theories depict these internal elements of human behavior in terms of concepts such as self, cognitive structure, ego functions in relation to id and superego, habit, family hierarchies, and personal constructs.

c. The external or *nutritional-social psychological* environment, prior to the mediation of cultural influences by parents, age-mates, and adult authority figures such as teachers, begins with the womb of the mother prior to birth. The Beadles have a highly readable chapter on "Man's double inheritance"— biological and cultural (1965, pp. 42–46)—in which they conclude, "We owe our evolution as much to our ability to develop, pass on, and modify patterns of behavior as to our physical inheritance as individuals," a view complemented by Bruner in his presidential address to the American Psychological Association (1965).

3. *Cultural patterns* prescribe what shall be learned and at what age different competencies are expected of girls and boys. Ferguson (1956, p. 129) supports this element of his conceptual framework by reference to a doctoral dissertation completed at McGill in 1955. Children reared in comparatively isolated environments displayed patterns of abilities which differed markedly from those of children reared in urban centers. For example, in the relatively isolated outport communities of Newfoundland, certain perceptual and motor abilities were highly developed, with verbal and reasoning abilities suffering in comparison. Moreover, there was considerable retardation in abstract thinking and concept formation. Clearly, the abilities of man are not culture-free; variations in environmental expectations lead to the development of different ability patterns. Apparently this statement does not contradict the views of contemporary students of behavior genetics who prefer "population" to "typological" thinking (Hirsch, 1967) and who are coming to appreciate an interactional approach (Glass, 1967).

4. *Abilities emerge through a process of differential transfer;* that is, change in one set of capabilities goes along with change in another (concomitant change). Suppose that x and y are per-

formance measures of two tasks which are believed to represent abilities, and t_x and t_y represent the amount of practice in them. Then a four-variable model for transfer can be written,

$$y = f(x, t_x, t_y),$$

which means that performance on one task is some unspecified function of performance on another task as well as the amount of practice on each. The simplest transfer function, $y = f(x)$, depicts concomitant changes in two measures of performance which are regarded as representations of underlying abilities. For example, we may observe a young child talking while playing and "placing" objects relative to one another: "This goes here and that goes there," looking to mother or to the bystander for confirmation. Later, the same child may learn, either in exploratory behavior or through guided discovery, to utilize the "placing" capability necessary to spatial relations in attaining a temporal concept, "This comes before that." Placing (spatial?) and verbal capabilities apparently facilitate the acquisition of further abilities.

Basing his generalizations upon Fleishman's studies of the factor structure of the learning task as practice continues (e.g., Fleishman and Hempel, 1954), Ferguson infers that "the abilities involved at one stage of learning differ from the abilities involved in another," thus supporting the hypothesis of differential transfer. In the Fleishman experiments, largely involving spatial, verbal, psychomotor, and habituated psychomotor response patterns, specific or "within task" factors sometimes accounted for a large proportion of the variance. For example, in a complex tracking performance, no more than 25 per cent of the variance could be explained in terms of identified ability factors. The progressive changes in the pattern of skill factors or abilities, which often seems to involve the emergence of specific capabilities,[3] may be a reason why psychometric prediction of trainability or job performance has had limited success.

[3] *Capability* has been employed in psychology to represent "the maximum effectiveness a person can attain with optimum training" (English and English, 1958, p. 1). In this book, however, we follow current usage (e.g., Gagné, 1962, 1965a, b) and infer the acquisition of response capabilities when a *change in performance* marks an individual's behavior in a given *stimulus situation* and the change persists over a period of time. The English word, *capability*, stems from the Latin root, *capax* and *capabilitas* (v. *capio*). Perhaps the shades of meaning may be conveyed by contrasting *non capax mentis* (not intellectually capable) with *non compos mentis* (not mentally composed) in which case the added dimension is the absence of smooth, effective functioning of the cognitive processes.

5. The concept of a general intellective factor, and the frequency with which substantial correlations are obtained among many psychological tests, can be explained by the process of *positive transfer*. Ferguson adds a corollary to his proposition that, through learning,

> [B]ehavior becomes organized, or structured, and to some extent predictable; [i.e.,] the distinctive abilities which emerge in the adult in any culture [are] those that tend to facilitate rather than inhibit each other. . . . Learning itself is viewed as a process whereby the abilities of man become differentiated, this process at any stage being facilitated by the abilities already possessed by the individual [1956, p. 121].

Neither in the early formulations of a theory of learning and human abilities mediated by differential transfer (1954, 1956), nor in his recent review article (1965), does Ferguson recognize the probability of facilitation by a set to learn (Harlow's "learning how to learn," 1949).

Hierarchical Organization of Knowledge

In a review chapter on "human abilities," however, Ferguson (1965, pp. 50–51) clearly is aware of Harlow's concept of learning sets and their significance for the issue of whether or not intelligence is fixed (Hunt, 1961, pp. 77–82). Moreover, he closely examines the "ingenious investigations" of Gagné and Paradise (1961), in which performance apparently depends more upon immediately subordinate learning sets than upon specific basic abilities. Ferguson goes on to illustrate the provocative lines of inquiry relating human abilities, learning, and transfer in the general area of research carried on by Gagné and his associates.

Gagné's hierarchical formulation of "the acquisition of knowledge" is based upon the notion of transfer from previously acquired, relevant learning sets (1962). The model may be useful for the conceptualization of emergent human abilities which are regarded as talents when they are socially valued. Beginning with subordinate capabilities (for example, symbol recognition, recognition of patterns, and number sense in the illustrative hierarchy, pp. 369–73), the reported experimental predictions and results obtained by Gagné and his co-workers would lead one to believe that learning to perform a given task becomes a matter of transfer from component learning sets to "a new activity which incorporates the previously acquired capabilities" (p. 364).[4] Like Ferguson, Gagné (1962, p. 363) is aware that an analogy may be drawn between his model and the findings of Fleishman and Hempel (1954) regarding motor tasks. Learning within a hierarchy seems to depend upon learning sets just previously acquired and not so much upon a basic factor or specific ability predictive of the ultimate performance.

The acquisition of knowledge. As stated earlier, Gagné (1962) has proposed employing a hierarchy of subordinate and coordinate capabilities, to be learned when necessary in order to master a superordinate set of capabilities which are required in order to perform a given task, acquire a terminal concept, or (by extension of the principles) to develop a compatible pattern of talented behavior. The proposal is based upon the notion of transfer from previously acquired, relevant learning sets (Harlow, 1949). The illustration diagrams a hierarchy of knowledge to be acquired for a "final task" of deriving certain mathematical formulae. The model specified the patterns of learning sets (subordinate capabilities) to be tested among ninth-grade boys who could not accomplish the final task. When the pattern of capabilities was identified, the model specified the learning program to be undertaken by each boy for mastery of the number series task. Gagné's book *The Conditions of Learning* (1965) employs versions of the hierarchical model to represent concept learning (p. 131), the learning of principles (pp. 150 and 155), and learning structure for number operations (p. 181).

Plans and the structure of behavior. Miller, Galanter, and Pribram (1960, p. 15) accept the evidence which supports the principle that behavior is organized simultaneously at several levels of complexity, and speak of this fact as "the hierarchical organization of behavior." Nevertheless, they deny that their notion has any relation to Hull's use of the phrase "habit-family hierarchy" and assert that they are talking about "a hierarchy of

---

[4] In his 1965 article for *The School Review*, Gagné draws a clear distinction between two different kinds of learned capabilities: *concept learning* (acquiring a common response to a class of objects varying in appearance) and *principle learning* (combining concepts into entities referred to as "ideas," "facts," and "rules" as well as "principles"). He also differentiates between *discovery learning* and *reception learning*, the active and passive modes, respectively, of learning principles. The two modes, and their ties to real or vicarious experiences, may represent the difference between "knowing that" (uncovering a principle for one's self) and "knowing about" (receiving information linking concepts together to form principles).

levels of representation." To guide behavior, a human being builds up internal representations which the authors term *Plans* (any hierarchical process in the organism that can control the order in which a sequence of operations is to be performed) made possible by the *Image* (the accumulated, organized knowledge an organism has about itself and its world). The hierarchy of TOTE (Test-Operate-Test-Exit) units which makes possible the execution of a Plan may be cast as an outline of operations or may take the form of a list structure in a program for a digital computer. Examples range all the way from the Logic Theorist (or LT program) employed by Newell, Simon, and Shaw (1958) to test ideas about human problem-solving behavior couched in terms of information processes, to the "mind-computer metaphor" recently reported by Colby (1967) in a paper on computer simulation of change in personal belief systems.

Images and plans (You imagine what your day is going to be and you make plans to cope with it) are mediating organizations of experience, operating on all levels simultaneously to explicate the manner in which behavior is controlled by an organism's internal representation of its universe (Miller, Galanter, and Pribram, 1960, pp. 6–15). The three self-styled "subjective behaviorists" (p. 211) suggested that linguistic analysis provides a model for the description of all kinds of behavior (p. 14). Consequently, they agreed upon assumptions (p. 18) which permitted them to explore the relationship between the *Image* and the *Plan*—the inferred multilevel molar units of their theory about behavior. Implicit in their approach is a new psychological principle; with guidance from the Image, higher-level Plans can be used to construct Plans which will guide behavior (or change the Image).[5] For example, the "grammar plan," with its hierarchy of grammatical rules of formation and transformation, operates to construct a motor plan tested for its "sentencehood" in speaking (pp. 139–58).

The Harvard-Penn-Stanford trio of psychologists believes that their multilevel, hierarchical Image-Plan conception of an internal representation escapes the kinds of criticism that have been leveled at constructs such as Tolman's "cognitive maps"

---

[5] Hebb (1960), in his appraisal of the behaviorist revolution in American psychology, employs the terms Metaplan (what Plan shall be in effect at any one time?) and Plan (determining the moment-to-moment course of behavior) in his argument that an analytical study of thought processes no longer can be postponed.

(1948), which took the form of inferred intervening variables anchored by independent and observable variables (1951, pp. 279–364). In his American Psychological Association presidential address, Hebb (1960) not only notes that the computer analogy (Plans) can include an autonomous central process as a factor in behavior, but also remarks that the representational "mediation hypothesis" (Osgood, 1953, pp. 392–412) is quite compatible with Woodworth's "schema-with-correction" in figure learning (1938, p. 74), since both are dependent upon contextual sensory cues. He also views the self as a complex mental process which the "tough-minded experimentalist in the problems of thought" should study analytically.

10

Human Talent
as Intelligent Behavior

Intelligent behavior, including that special dimension we know as "talent," is probably a complex set of phenomena, influenced by variables which are not yet thoroughly understood in spite of our continuing technological sophistication. There are at least two possible causes for this lack of understanding: First, each technological development totally depends upon the intelligent behavior of the developers, and second, each of our measures is dependent upon the "thing" measured, rather than independent of it, as is usual in scientific measurement. This argument brings out a serious logical flaw, which may necessitate some reevaluation and redirection of the measurement of intelligent behavior; thus, research efforts may need to take new and as yet unconceived directions. Often measures of intelligent behavior are thought of as some sort of intelligence test. When pressed for a definition of intelligence, Jensen (1969) was forced to define intelligence to be what intelligence tests measure, an unfortu-

Adapted and excerpted from C. McGuire, *The Years of Transformation*, Chapter II, Final Report, Project Number 742, Contract Number 5–0429, to the U.S. Department of Health, Education and Welfare Office of Education, Bureau of Research.

113

nately necessary tactic whether or not one agrees with Jensen's positions and interpretations. A logical inconsistency of this magnitude, that is, a case of the definition's necessarily including the object of definition, highlights some of the inadequacies which stand in the way of a more thorough understanding of the development of intelligent behavior. In a symposium presented at the 77th Annual convention of the American Psychological Association, the participants were reluctant to state a definition of their subject matter, possibly in part because of a lack of consensus (Rowland et al., 1969). The definition proposed which appeared to be least objectionable to the majority of participants was that *intelligent behavior is the development of a behavioral repertoire which allows the individual to function effectively and efficiently in his or her particular environments.* This is, of course, a global definition which will need scientific interpretation before independent and/or cooperative research studies can be carried out or evaluated in light of this approach. The hopeful aspect of the current state of this area of psychology is that many of the research studies which have been carried out, such as those reported here, complement one another and seem to be leading to the development of a convergent metatheoretical understanding of intelligent behavior.

A trend among many psychologists interested in human intelligent behavior is to place less emphasis on the classical laboratory experiment with the manipulation of a limited number of variables. One alternative seems to be the *natural experiments* in which relevant elements of the context may be taken into account. When the researcher works from models (which should never be reified, but rather, should be constantly improved with advancing knowledge), he can introduce new variables not only by means of direct operations and measures, but also by selecting sample populations and by employing multivariate and often indirect evaluations made possible by the use of high-speed computers. In order to proceed in this manner, we must grant that human learning concerns something more than basic habit-formation and allied reflexive forms of behavior. Naturalistic research may possess inherent values. The inquiries and the ideas generated from this kind of research, can be used to approach complex human learning situations and the interplay of factors which enter into the processes of acquiring skills and knowledge, orientations, and behavior patterns.

Dilemmas of a Developmental Approach to the Study of Talented Behavior

In 1960, the *American Psychologist* published a Bingham Memorial Lecture by Dael Wolfe, advocating the development of a diversity of talent. Consequently, one might readily assume that the study of talent, i.e., valued behavior, would be an acceptable focus for research in child and human development. Such is not the case unfortunately; in the behavioral sciences of the mid-twentieth century, the study of talent has become an ambiguous area of inquiry, whereas there has been a continuous study of the nature of "intelligence," as the abundant literature testifies. The behavioral scientist who sets out to formulate reasonable schemas, and to test propositions which follow from a conception of human intelligent behaviors as products of developmental changes, has to resolve a number of dilemmas which currently block communication.

First, there are the residuals of the nature-nurture controversy of the early 1930's, which have been recently restimulated by Jensen (1969), and by recent distortions of the concept of the genetic contribution to intelligent behavior as seen in the work of Garrett (n.d.). In the early decades of this century, the prevailing concept held by psychologists, social workers, and educators (as well as laymen and members of the medical profession) was that intelligence was a fixed, genetically determined individual characteristic. One corollary was that abilities (what one "can" do) are biologically predetermined, not culturally influenced, and that the ability to learn depends largely on maturational readiness. As indicated earlier, Hunt's *Intelligence and Experience* (1961) is a comprehensive summary of evidence contradicting such beliefs. Although many educated people still cling to the assumptions of fixed intelligence and predetermined development, these ideas no longer are the accepted premises underlying a majority of the decisions about the educational process in the United States. There are, however, glaring exceptions to this development, such as the public school system of New York City, where an outmoded practice of "tracking" a student in the early elementary school years is still in vogue.

IDENTIFICATION VERSUS DEVELOPMENT

A second dilemma stems from the existence of two paradigms which shape conceptions of thought and implicit assumptions about the nature of human talent(s): the *mineral model* versus the *agricultural model*. Most of the English-speaking world has employed the mineral or mining approach, emphasizing the identification of talented behavior and the implication that "talents" are to some degree inherited or at least inborn. To illustrate, within the past decade, the Social Science Research Council had received a report from its Committee on Identification of Talent (McClelland et al., 1958). Neither that report, nor John W. Gardner in his discussion of "the search for talent" in his treatise on *Excellence* (1961, pp. 46–53), accepted the premise of inherited talents. For example, Gardner writes, "I am concerned with the social context in which excellence may survive or be smothered" (p. xiii). McClelland and his associates begin their volume (pp. 1–28) with a discussion of issues in "the identification of talent" but, in the concluding chapter, assert that "basically ability refers to the adaptiveness of behavior" (p. 235) and that "the 'talent' is in the combinations of a particular person with a particular situation" (p. 236).

The alternative view, a developmental approach, naturally suggests an emphasis on the cultivation of talented behavior during the early years of schooling, provided that there is an appropriate cultural milieu. The U.S.S.R. has had the necessary climate for such a view of human talent to take effect. In the Soviet *Weltanschauung*, by official decree the belief is that there are no inherited intellectual differences among individuals. Officially, individual differences are recognized only where there is brain damage or a comparable insult to the otherwise intact human organism which brings about malfunctioning. This belief in innate equality leads to an explanation of differences in performance by attributing them to motivational factors as well as to inequalities in prior experiences. According to observers, the approach seems to be effective. For example, Bronfenbrenner (1962) reported after a visit to the Soviet Union that the assessment practice of Soviet educators, according to which they evaluate the performance of each child on the basis of how well his group does as a whole, appears to have positive outcomes. Moreover, the ones who learn most readily in the groups (which are formed in elementary classrooms on a random basis) apparently spend a great deal of time and energy helping other members. Thus they act to maximize group performance and thereby their own grades. Nevertheless, Bronfenbrenner believes that, in the future, Soviet students of human behavior are going to pay less attention to conscious process (p. 82) and more to studying the process of socialization as well as mechanisms of social control. The impetus, he forecasts, will come from increased concern about providing "the most favorable conditions for the education and communist upbringing of the rising generation" (p. 80).

NATIONAL POLICIES AND
LINGUISTIC CONVENTIONS: ASSUMPTIONS

The ambiguity of approaching human talent(s) as either a matter of identification or cultivation becomes even more complex when ideologies and national policies are considered. In the Soviet Union, as stated, the official point of view is that inherent differences among individuals with intact organisms are non-existent. For all intents and purposes, the view was made official by decree on July 4, 1936. Brozek, in his thorough interpretive review of developments in Soviet psychology (1964), implies that a process of "de-Stalinization" has been taking place in Soviet scientific as well as political life. Consequently, Soviet-psychologists are beginning to criticize their colleagues for too narrow an interpretation of the 1936 decree against "tests," an interpretation which resulted in a neglect of the study of individual differences among children. One of the casualties appears to have been Vygotsky's seminal 1934 monograph on *Thought and Language* which has been translated (1962) and made available in English in a paperback edition. This monograph is referred to infrequently by Russian scholars.

The assumptions which underlie American education—that man is potentially "good" and that this "good" can be brought about by equality of educational opportunity—represent a faith in the principle of the perfectibility of man. This belief in turn "implies the ability of all to learn, and the duty of society to teach" (Keppel, 1966, p. 11). In order to maintain these ideals, and in order for some to reconcile their mode of thought with the emergent *zeitgeist* in psychology and education, English-speaking peoples have employed at least two convenient fictions: the concept of *alpha-* as well as *beta-intelligence,* and the substitution of equal treatment for equal opportunity.[1]

Convenient fictions of the English-speaking people. The fictions about intelligence and equality stem from customary ways of representing thoughts, feelings, values, and actions in the past. The two sets of concepts have become interrelated over the years, and relatively few persons stop to examine the assumptions implicit in the language usages which carry over from earlier world-views. With reference to *alpha-* and *beta-intelligence,* Donaldson (1963, pp. 6–7) inadvertently but succinctly demonstrates the difference between *potential* intelligence (A) and *realized* intelligence (B). She shows that the ambiguity of the English "he cannot swim" is clearly differentiated in the French *il ne peut pas nager* (because of lack of potential, a radical incapacity) from *il ne sait pas nager* (because he has never learned, never realized his potential for achievement of the skill). The custom of associating "innate" or "native" with "potential" and the retention of language usages from the past without reflection upon implicit assumptions are illustrated in a form from a graduate school requesting evaluation of "native intellectual ability" of a student seeking a grant-in-aid to continue his education. The evaluator usually realizes he is being asked to make a judgment about probable quality of future intellectual performance and responds accordingly. The substitution of equal treatment for equal opportunity has been depicted most succinctly in *Who Shall be Educated* (Warner, Havighurst, and Loeb, 1944), especially in the chapter entitled "Curricula—Selective Pathways to Success." They comment somewhat ironically on the reaction of people in Hometown against a differentiated curriculum in the high school: "The democratic way seems to be to give everyone the same educational opportunities—the same as required for those at the top, namely, college preparatory courses" (p. 69). The three Chicago behavioral scientists recognized the potentialities present in emerging high

---

[1] After formulating the concept of convenient fictions used by English-speaking peoples, independent presentations of each idea were encountered, but the authors did not interrelate them as part of a linguistic *Weltanschauung,* or perception of reality associated with a language. Donaldson (1963, pp. 1–9) argues "we cannot prove that the development of intelligence is a process of inevitable unfolding" yet we differentiate between Intelligence A (innate capacity) and Intelligence B (developed intellectual power). Komisar (1966) examines the "two faces of equality," then argues the specificity of equal treatment and the generality of equal opportunity to "solve" the seeming paradox of equality in schooling.

schools of the "comprehensive" type some years prior to the Conant report upon the American high school today (1959 and 1964, pp. 21-47). In such schools, all students have a common "core curriculum" (usually English, social studies, physical and health education, perhaps some form of mathematics, and sometimes a science) supplemented by individual electives, but there are no hard-and-fast divisions into college preparatory, business, and other programs.

Language use: growth and/or development? One approach to the study of change over time has been to assume that "growth" is subsumed in the concept of "development." Furthermore, the use of the two terms together, as in "human growth and development" (often encountered in education), in this context would involve an unnecessary redundancy. In a somewhat different approach to discussing the nature of cognitive development, Bruner apparently elects to use "growth" to represent transformation to a more developed or mature stage; for example, "The Course of Cognitive Growth" (1964), "The Growth of Mind" (1965), and a book with collaborators, *Studies in Cognitive Growth* (1966). Nevertheless, in the preface to *Studies,* Bruner describes the focus of work at the Center for Cognitive Studies at Harvard University as "exploring the course of cognitive development" (1966). Parenthetically, in their *Dictionary,* English and English (1958) point out, "Originally *development,* as a qualitative phenomenon, was distinguished from *growth* as quantitative or incremental." They continue "present usage tends to make development inclusive of growth or to employ them synonymously" (p. 148). They note that the distinction is not well observed and indicate a preference for the term "development" to refer to "change toward a more developed or mature state" (p. 233) instead of "growth" which has no adjectival form, as in "developmental psychology."

Talented Behavior as Behavioral Capabilities

The construct of an "ability" may range from overlearned patterns of behavior—e.g., verbal or motor abilities—which tend to vary from one culture or subculture to another, through task (or job) performances that require the emergence of specific capabilities, to the acquisition of knowledge—i.e., cognitive abilities, structures, or schemata). Any distinction drawn among cognitive, instrumental, and motor abilities would appear to be arbitrary and based upon classifications of the tests employed as measures of criteria. The common element in human abilities seems to be behavioral capabilities which appear to be subordinate, coordinate, and superordinate to one another in the hierarchies that make up talented behavior. The subordinate capabilities may range from relevant learning sets such as symbol, pattern recognition, or closure (and number sense in Gagné's model), through "advanced organizers" which apparently facilitate integration of new material into existing "cognitive structure" (Ausubel, 1960, 1963), to the biological or physical capabilities necessary for specific kinds of athletic, musical, and artistic talent.

MODELS FOR SETS OF BEHAVIORAL CAPABILITIES

To explain the ways in which factor analysis can contribute to psychological theory, Guilford (1961) described three models for representing ways of functioning within and among individuals. As Allport does, (1937, 1966), Guilford refers to "traits" (inclusive of abilities) as relatively enduring ways in which one person differs from others. There appear to be three major models for representing behavioral capabilities:

1. *Dimensional model.* To the extent that an individual's ways of functioning can be accounted for in terms of a limited number of common factors, each person can be represented by a point in n-dimensional space. Guilford's example (1961, Fig. 1) is a dimensional model with three axes which have as their point of origin (intersection) the center of a sphere. The axes are linear dimensions representing a unique trait, or common factor, along which individuals have characteristic positions reflecting individual differences. For example, individuals P and Q are each described quantitatively by their projections upon the three axes. The orthogonal projections define points which are each person's characteristic position in this particular three-dimensional space. Guilford is quite aware of intra-individual differences from time to time and, consequently, uses the term "characteristic position." He suggests that persons who find difficulty in conceiving of a space with a large number of dimensions (hyperspace) should think instead of a profile chart in which the dimensions are laid side by side.

2. *Hierarchical model.* A hierarchy involves the arrangement of elements into a graded series, orders, or ranks, such that each element is subordinate to the one above and coordinate with those at the same level. Sir Cyril Burt (1940, 1949), who began his studies two decades earlier, was the first to advance a hierarchical group-factor theory of the structure of human abilities. In his 1950 book, Vernon employed Burt's four-level model (p. 22, Fig. 2) for his hierarchical approximation of mental structure, which was composed of a general intellectual factor (g) superordinate to two major group factors, the verbal-educational (v:ed) and spatial-practical-mechanical (k:m) groups. In a recent Bingham Lecture on "Ability Factors and Environmental Influences," however, Vernon (1965) concluded that "there is no one final structure, since so much depends on the population tested, its heterogeneity and educational background, the particular tests chosen, and the techniques of factorization and rotation employed." Nevertheless, the English educational psychologist retains the general intellectual factor as well as his v:ed and l:m group factors in his diagram of "the main general and group factors underlying tests relevant to educational and vocational achievements." Guilford (1961), however, goes back to his book on Personality (1959a) for a hierarchical model "treeing" down from two syndrome types (for example, strength of character and general self-restraint) linked by one of the primary traits (for example, honesty), each of which has subordinate traits at the "hexis" level (for example, impulse control). Each of these habit-patterns traits (third level) is grounded in fourth-level specific-action traits (for example, resistance to cheating observed in a series of opportunities).

3. *The matrix model.* To represent the third model which "comes about from attempts to discern logical relationships among known factors," Guilford (1961) elects to illustrate the matrix with his "cubical model of the structure of intellect," representing categories of primary abilities with respect to three modes of variation (1959b). The three-dimensional matrix has five intellectual operations crosscutting four kinds of content, yielding six types of products, that is, 120 cells or "primary mental abilities instead of the 55 presently recognized" in Guilford's words (1961). Humphreys (1962), in a thoughtful paper on "The Organization of Human Abilities," grants that test behavior can be made constantly more specific but does *not* believe that the 120 test behaviors suggested by Guilford's structure-of-

the-intellect model should be regarded as definitions of "primary" factors. Rather, Humphreys moves on to Guttman's facet theory (1958) which he seems to regard as a matrix-type model for the definition of a universe of possible tests.

FACET DESIGN AND TEST CONSTRUCTION

Ferguson (1965) does not agree that the facet model, which Humphreys illustrates by a three-dimensional, Guilford-type block, is a more general substitute for the hierarchical model. As Ferguson points out, a matrix model is merely a descriptive statement of the organization of human abilities at a particular point in time and does not deal with the developing and changing structure of abilities (or talents) in the child and/or adolescent (1965, p. 48). Nevertheless, Humphreys' suggestion that facet theory would facilitate controlled heterogeneity in test construction appears to have much merit. For example, a sixty-item reasoning test would be defined by the item-content facet (use of numbers, words, figures, and photographs), the item-format facet (use of analogies, series, and classification), and the categories-of-reasoning facet (numerical, mechanical, abstract, inferential, and intuitive). Humphreys (1962) asserts that according to facet theory, there are no "pure" tests; thus, controlled heterogeneity should be the goal of test construction. The most recent presentation of Guttman's facet theory (1954) may be found in Foa's account of facet analysis and design (1965), in which interpersonal behavior is defined as the Cartesian product of the observer by perceptual and behavioral facets. Foa also suggests that experiments employed by Bruner, Goodnow, and Austin (1956) in their studies of the thought processes could be conceptualized in terms of facet theory.

Forms of the hierarchical model for human behavior. Modern behavioristic conceptions of a symbolic or internal mediating mechanism to represent the past in the present tend to parallel Hull's principle of the habit family hierarchy (1952, pp. 256–74) or motor equivalence governing behavior in space. Mechanisms invoked to explain sequences of symbolic responses or trains of thought also borrow his notion of divergent and convergent mechanisms as well as the concept of a fractional anticipatory goal response (Maltzman, 1965; Staats, 1961). Most of the theorists, however, are quite aware that Hull's "hierarchy" reflected observed external spatial behavior of animals rather than internal thought processes of human subjects. Hull's initial formulation (1934) reflected an ordering of alternative (interchangeable) responses ranked according to their strengths. Hull's hierarchies, though based on observed spatial behavior, can be applied by logical analogy to cognitive behavior. By analogy, cognitive behavior conceptualized in a hierarchical structure would be an ordering of learned alternative intellectual behaviors.

The most familiar hierarchical models place capabilities (factors, operations) in a subordinate-coordinate-superordinate order proceeding from the more specific to the more general. In his 1965 Bingham Lecture, in which he pinpoints environmental influences which underlie the development of different patterns of human abilities, Vernon employs a hierarchical model. He uses the model to simplify the problem of designing cross-cultural studies in the emergent, nontechnological nations of the world, with the intention of cultivating the potential talents of their young people. The illustrative hierarchical model chosen by Guilford (1961) looks somewhat like a genealogical tree with four levels upward from specific actions to two personality syndromes (strength of character and general self-restraint). The syndromes have in common a number of acts of resistance and habit patterns as well as a primary trait (honesty). In a recent proposal for an ability-oriented concept of personality, Wallace (1966) questions some of the assumptions underlying essence conceptions of personality, which typically involve hierarchical arrangements of response predispositions (e.g., needs or traits) from a sample of responses (to ambiguous projective materials, for example). He would reduce stimulus ambiguity to determine whether or not given individuals "are simply capable or incapable of certain response in certain stimulus situations" (p. 133). When the proposal is put to a test, however, there is a likelihood that the repertoire of observed response capabilities may turn out to be arranged hierarchically in a superordinate, coordinate, and subordinate manner with their occurrence dependent not only upon specific stimulus situations, but also upon symbolic transformations of perceived cues.

CENTRAL REPRESENTATIONS
IN RELATION TO OBSERVED BEHAVIOR

For the most part, the use of constructs such as plans and metaplans, images, cognitive maps, mediation processes, and schema permits us to discuss behavior in molar terms without specifying the precise biological substrates of the processes postulated to underlie behavior observed in response to independent stimulus situations. For example, Hebb (1949) recognized that phenomena labelled set, attention, attitude, expectancy, hypothesis, intention, vector, need, perseveration, and the like have one element, and one only. The influence is not an immediately preceding sensory stimulation for each response but an ongoing central activity which Hebb labels the *autonomous central process* (p. 5). Hebb concluded, "The problem for psychology then is to find conceptions for dealing with such complexities of central neuron action: conceptions that will be valid physiologically and at the same time 'molar' enough to be useful in the analysis of behavior" (p. 11). His theory of brain function, suggesting that the thought process involves a phase sequence composed of a series of cell assemblies ordered in time, was published during the same year (1949) as Harlow's report of investigations into "learning to learn." Harlow's "learning sets" were regarded as analogues of strategies for information processing acquired by rhesus monkeys from experiences with a given kind of problem. Four years later, after work with human subjects in addition to his primates, Harlow pointed out some serious limitations of a drive-reduction theory of motivation (1953).

During the next year, Hebb addressed experimental psychologists upon the notion of drive and the conceptual nervous system (C.N.S.). He adopted Bergmann's view (1953) that "intervening variables" and "hypothetical constructs" are functionally similar. Consequently both intervening variables and hypothetical constructs can properly appear in the same theory. Hebb's published paper (1955) referred to a range of studies to illustrate the "cue" and "arousal" functions of the nonspecific projection system.[2] In alertness, emotionality, and curiosity. Linking psychological with physiological terms, he urged research on

[2] Ascending and descending projective fibers link the cerebral cortex with the RAS or recticular activation system which provided a referent for Lindsley's "activation theory of emotion" (in Stevens, 1951, pp. 473–516). They also are linked with the limbic system that borders the hypothalmus and which is strategically located for "the correlation of feelings, particularly those arising from internal organs of the body" (the Papez-MacLean Theory of emotion, Morgan, 1965, pp. 311–12). Thus the RAS and limbic systems, both associated with the hypothalmus, mediate emotional experience and expression.

the "immediate drive value" of "cognitive processes" which, without an intermediary, provide cortical feedback to the arousal system. As indicated in a seminal article, Hebb (1960) contended that the first stage of a revolution in psychological thought (explaining simpler behavior through an experimental application of the S-R paradigm) had been completed; and now, the time had come for an attack upon more complex behavior, particularly the thought processes, while maintaining liaison (translatability of terms) among different universes of discourse.

DYADIC INTERACTION

Sears (1951) laid a foundation for a convergent theory of intelligent behavior [3] in his presidential address to the American Psychological Association. Sears viewed a theory as "a set of variables and the propositions that relate them to one another as antecedents and consequences," wherein intervening variables ultimately have to be reducible to operations. He recognized two advantages of theory development: first, relationships observed in research have greater generality if the variables involved are part of a more global concept; and second, a worthwhile theory permits the use of multiple variables together with their relating principles, in various combinations for the prediction of events as well as the study of changes over time.[4] Moreover, Sears' current view (in Stevenson, 1966, pp. 36–39) is that "personality" and "social psychology" are but one field of study as implicit in his 1951 statement. Pointing out that psychologists tend to think and theorize monadically ("they choose the behavior of one person as their subject matter"), he made a strong case for a *dyadic* unit of behavior, "one that describes the combined actions of two or more persons," as the locus of theory and research in human behavior. In general, motives, habits, cognitive capacities, ego organization, and even "on-going action" are shaped by an individual's *expectancies* that others will demonstrate toward him either supportive or nonsupportive behavior. In Sears' schematic representations, the individual's

[3] A schematic diagram of relations among variables involved in a dyadic interaction context model is proposed in Chapter 11 of this book (p. 138). The model employs Sears' terminology and symbols in representing a theory of human development and intelligent behavior.

[4] Principles of theory building, the reduction of intervening variables to operations, and the testing of hypotheses are illustrated in Chapter IV of *The Years of Transformation* (McGuire, 1969).

potentialities for action are specified by motivation and cognitive structures which are largely a product of learning, that is, changes in potentialities for action.

By the middle of the twentieth century, then, Hebb, and Tolman had introduced the notion that central processes govern behavior; Harlow not only had established the concept of "learning how to learn" but also had pointed out limitations of a drive-reduction theory of motivation; and Sears had shown that all human behavior is inherently social. Furthermore, in complementary theory and supporting research, other behavioral scientists (Erikson, 1950; Havighurst and Taba, 1949; Hollingshead, 1949; Kluckhohn and Murray, 1953; Lerner and Lasswell, 1951; Parsons and Shils, 1951; Warner et al., 1949) had established meaningful linkages among cultural contexts, social structures, personality, intellectual functioning, and observed behavior, together with "markers" of prior experiences in family backgrounds (Warner, Meeker, and Eels, 1949) and in age-mate societies (Clark and McGuire, 1952; McGuire and Clark, 1952). In retrospect, one would infer that the midcentury nexus of developments in psychological, anthropological, and sociological theory and research, together with the contributions of the individual scientists involved, was primarily responsible for the launching of the current *zeitgeist* in the behavioral sciences and the developing renaissance in education.

Intelligent and Reflexive Behaviors

The word *intelligence* is derived from the Latin words *intus legere*, meaning "to read what is within" and implying some internal or central process. Behavior governed by such a central process or operation would then be understood as "intelligent." [5]

---

[5] Piaget (1966, pp. 3–17) has discussed the adualistic (biological and psychological facets) and adaptive nature of intelligent behavior and, in a significant article (1961), he considers three interacting processes or factors which underlie formation of a system of dynamic transformations that permit thought and operations which make us understand the state of things: (1) maturation (the invariant sequence of changes, ordered but not age-related), (2) physical and logicomathematical experiences (involving learning through the organization of experience as well as equilibration—assimilation or the incorporation of objects into patterns of behavior and concurrent accommodation of schemata), and (3) social interaction (reciprocal stimulation in dyadic and polyadic transactional systems which foster attainment and maintenance of a dynamic equilibrium through a train of exposures to conditions and events, i.e., adaptation).

In contrast to intelligent behavior is sense-dominated behavior (Hebb, 1966a, p. 83), cue-dependent behavior (Miller and Dollard, 1941, pp. 21–28), or stimulus-bound behavior which, with certain reservations might well be designated as *reflexive behavior*.

Reflexive behavior shows a close temporal relation between stimulus and response, assuming that it depends upon straight-through connections in the C.N.S. (Hebb, 1966, p. 83), and tends not to function in a mediated or hypothesis-testing manner.

Hebb (1966, pp. 82–83), in a discussion of behavior, uses a spider spinning a web and a man ploughing a field as his examples. Both are seeking to obtain food, but one is planning and the other's search appears aimless. The important difference is that some behaviors show a close temporal relationship between the stimulus and the response. The type of behavior demonstrated by the web-making spider is cue-dependent or sense-dominated, and within the proposed frame of reference, such behavior would be termed *reflexive*. The other behavior depends upon mediating processes, such as ideas and thinking, and would fall into the classification of *intelligent* behavior. As Hebb points out, behavior is fundamentally an adaptation to the environments; therefore, both reflexive [6] and intelligent behaviors need the informational input from the senses, i.e., all behaviors are affected by sensory feedback at all times. The distinctive criterion is the mediational aspect; the higher the behavior the more sym-

---

[6] Berlyne (1965, p. 11) observes that the word "reflex" now is virtually never used by Western psychologists "except to denote the rigid, unlearned behavior patterns that make up a relatively small part of higher mammalian behavior." Russian psychologists use "reflex" to designate a stimulus-response association which is highly susceptible to modification by central processes. Much of the current research, particularly that of Anokhin (in Cole and Maltzman, 1969) on "Cybernetics and the Integrative Activity of the Brain" (pp. 840–56) given at the 1966 International Congress of Psychology in Moscow, is in accord with the Pavlov-Sechenov tradition. Apparently Berlyne thinks Russian "reflexive (*reflektorny*)" point of view is to be clearly distinguished from "stimulus-response" psychology. According to Anokhin, in "Ivan P. Pavlov and Psychology" (Chapter 7 in Wolman, 1968, pp. 131–59), Pavlov observed the manner in which the conditioned reaction precipitated "signal" activity of the central nervous system. This "warning" character reflex enables the organism to adapt itself to events which are not yet taking place (foreseeing, predicting). Pavlov put forward the idea of a "dynamic stereotype" that structures the situation in which the conditioned reflex acts. Pavlov and those who followed him seemed to believe that the "second signal system," involving verbal stimuli and verbal responses (reciprocal stimulation), being peculiar to human behavior, gives rise to behavior which in far-reaching respects differs from that produced by the conditioned reflexes of the "first signal system."

bolic mediation there tends to be. However, even highly logical behaviors require sensory guidance to permit evaluation and maintain schema-with-correction. Therefore, one may say that behavior may be reflexive without being intelligent, but behavior may not be intelligent without some aspect of sensibility and/or the use of acquired reflexive capabilities.

Within the behavioral realm, the intelligently behaving organism may be said to adapt to or to manipulate the environment, often by combining learned reflexive capabilities to cope with new situations and events (usually in combinations not previously employed), in order to make necessary adaptation more probable. Adaptation is understood to be an active process which is stimulated or inhibited by feedback to the organism from one or a combination of the three environments. This feedback would be positive (confirming or continuous) or negative (nonconfirming or discontinuous) according to the meanings attached to the environmental symbols and objects, or to the degree of familiarity existing between the learner and the object of learning.

Environmental symbols and reciprocal interaction. An environmental symbol or object may take any form, for the intelligent organism's perception of that form and the attachment of meaning to the perceived form control the nature and degree of adaptation. This may be understood as an evaluative process, and as such would be most relevant to an interpretation of the behaviors of the intelligent organism's transactions in a social situation. The intelligently behaving organism interacts with objects and *cultural agents* who are significant figures of the external environment. The intelligently behaving organism (the learner) may himself be a cultural agent inasmuch as he establishes goals or expectations for evaluating his own behavior. The learner compares his behaviors to the expectations of others in two ways: First, he responds to certain cultural and social expectations imposed by the cultural agents, and second, he behaves in terms of his own expectations about the probable supportive or nonsupportive behavior of the cultural agent (usually in terms of acceptance-rejection and/or approval-disapproval) in response to the learner's doing and/or being in proximity to the cultural agent.

In order to avoid reification of these models as entities (a tendency when one employs nominal terms), we give the three forms of behavior the status of multidimensional attributes. Each form of behavior is subject to the processes of evaluation, either by cultural agents or by one's self, and to reciprocal stimulation—a dyadic and/or polyadic phenomenon analyzed at its beginning during infancy in terms of four developmental principles set forth by Harriet Rheingold (pp. 1–17) and discussed by Sears (pp. 36–39) in Stevenson's seminal monograph, "Concept of Development" (1966).

If there is *continuity* in the form of confirming feedback between the behavior and self/other expectations, no adaptation (*secondary behavior* in cybernetic terms, or *accommodation* in Piagetian terms) is necessary. Should there be *discontinuity* between the behavior and self/other expectations, adaptation is necessary to the degree that the learner has attached meaning to these environmental symbols.

Definition of intelligent behavior.   Intelligent behavior cannot be understood as an S-R connection, or as response to external stimuli. Such behavior would be simply reflexive. If an intelligently behaving organism is capable of attending to stimuli selected from its genetic, internal, or external environments, then the level of intelligent behavior may be measured or evaluated most effectively according to the level of adaptation (including the use of previously acquired reflexive capabilities) in relation to the degree of environmental complexity. This statement means that the higher the degree of environmental complexity to which the organism can adapt successfully, the higher the degree of intelligent behavior.

The definition of intelligent behavior may be summarized as an adaptive act or sequence of responses controlled by a central process which responds according to the learned meaning attached to environmental symbols or objects and events encountered. From this definition, three principal factors or concerns may be extracted:

1. The nature of a central process which controls, guides, or directs behavior
2. The nature of adaptation
3. The method by means of which meaning is attached to environmental symbols, objects, or events

The final factor appears to have been the dominant concern of earlier experimental psychologists.

Learning as organization of experience.   Experience is the relationship of familiarity between the organism and its three interacting environments: biological heredity, one's self, and a social heritage (represented by cultural agents). In the theory developed, "experience" brings about transformations in the organization (or schemata) of central processes (Berlyne, 1965, pp. 113–23). The "equilibration," to use Piaget's term (1961), is brought about by encounters with discontinuities in one or a combination of any of the three interacting environments, which form the nexus of "being human."

The nature of self. In terms of the context of dyadic interaction, a self or personality becomes an important facet of the internal environment as a consequence of the reciprocal stimulation between the developing organism and cultural agents who are objects of identification in the external environment. The most significant cultural agents who provide models in social learning are parents (closely tied authority figures with emotional involvement in the new "individual replacement"), more remote adult authority figures such as teachers, and age-mates who can accept, avoid, reject, or isolate the new member of a human society.

Translation to a research model. The dyadic model of intelligent behavior may be translated into a research model in which the behavior (B) to be explained or predicted is set forth as a function of antecedent underlying potential capabilities (P_a) assessed at some prior time, expectations or attitudes ($E_{a.b}$) assessed in the learner with reference to the cultural agents, and the probable evaluations of the learner expressed in terms of antecedent responses ($R_{b.a}$) of cultural agents to the learner (McGuire, 1961).

Theory and Criteria of Talented Behavior

The explanation and prediction of various kinds of talented behavior requires (1) an underlying theory or model such as the one set forth here, (2) a set of relevant criterion measures, and (3) suitable methods for mapping out and combining the promising indicators or dimensions of what is being studied. A schematic diagram of a model for research in human talent has been set forth (McGuire, 1969) previously. Essentially, a significant portion of the observed and valued behavior of an individual may be represented in the form of an equation whose terms can be identified empirically and, when confirmed, can be transformed to a linear mathematical model for necessary computations.

$$B_a = f(P_a, E_{a.b}, R_{b.a}), S_a, G_{a.b}, C_{ab}$$

The model proposes that various kinds of talented or intelligent behavior, assessed by criterion measures, are a function of several kinds of variables. One set is within the person (a) but is influenced by cultural agents (b) such as parents, peers, and teachers. These variables involve first, potential cognitive, perceptual, and other abilities as well as the deeper elements of personality (P_a); second, expectations about one's own behavior and the probable responses of significant others, often expressed as attitudes ($E_{a.b}$); and third, responses of other persons to the individual concerned, experienced in terms of pressures imposed upon the one observed ($R_{b.a}$). At least three kinds of modifying influences have to be considered. One is the sex-role identification of the individual and the sex-typing of socialization pressures upon him or her (S_a). Another, not included in earlier paradigms, is the generation or age-mate society to which the person belongs or refers and in which others react to the one being observed ($G_{a.b}$). Finally, there is the situation, or context of behavior (C_{ab}), such as a community or school setting which provides an institutional framework along with certain learning opportunities and impersonal expectations (or the setting in which a natural or a laboratory experiment takes place).

11

Toward a Theory
of Intelligent Behavior:
A Proposed Model

Education is seen as a behavioral science consisting of planned intervention into developmental sequence. Intelligent behavior is the objective of education; the purpose is behavior modification toward socially determined criteria. With a dyadic model, intelligent behavior is shown to be the result of the invariant processes of (1) reciprocal interaction between the organism and its genetic, internal, and external environments, (2) the acquisition of experience which involves transformation within the organism, and (3) the development of central processes of control. The operational principle is seen as continuity-discontinuity in feedback from the organism-environment interaction.

The development of intelligent behavior, especially as it is influenced by school practices, is the focus of the emerging

Adapted from "Toward a Theory of Intelligent Behavior: A Proposed Model," *Journal of School Psychology*, VII, No. 4 (1968–69), 50–56. This chapter is based on a paper presented at the 15th Annual Convention of the Southwestern Psychological Association, New Orleans, April, 1968. It was initiated and sustained at times by grants from the Hogg Foundation of The University of Texas. The authors' conceptions of "Education as Behavioral Science" were formulated while engaged in research and development activities under the auspices of the Research and Development Center for Teacher Education, University of Texas at Austin, Contract Number OE 6-10-108, sponsored by the University in cooperation with the United States Office of Education.

discipline of educational psychology. A model to explain intelligent behavior from a developmental frame of reference is proposed in this chapter. This dyadic model is based upon the concepts of reciprocal stimulation and the social and sequential nature of human development. An invariant sequence in the development of intelligent behavior implies a continuum with little specificity other than a beginning (conception) and an end (death). To relate any specific point on this developmental continuum with a chronological age would be to misconceive the essential notion of sequence. In the development of intelligent behavior, time is a variable of little significance; instead, it is the background against which the developmental panorama transpires.

A Behavioral View of Education

For the behavioral scientist, education involves planned intervention into human development.[1] The intervention usually requires interaction and the beginning of some form of interpersonal relatedness, wherein the interaction produces experience and the learner extends his control over self and the environment. Therefore, educational encounters necessitate interaction and inevitably allow the acquisition of experiences which become a part of the central processes guiding subsequent teaching and learning behavior (i.e., intelligent behavior). The notion of education as a behavioral science converges from the genetic epistemology of Jean Piaget and the cognitive functionalism of Jerome S. Bruner and Irving Sigel.

Piaget's Psychophilosophical Conceptions

The most elaborate attempts to come to terms with the conception of an invariant sequence in human development are known as genetic epistemologies, a term originated by Baldwin

---

[1] Planned interventions usually involve "symbolic interaction" (Blumer, 1962) and take the form of educational encounters between a learner (Alpha) and cultural agents (Betas). Although the emphasis herein is on preparing teachers to interact with pupils in educational encounters, cultural agents may be a counselor, special educator, computer interface, books in a classroom or library setting, age-mates, and sometimes one's self on a videotape or a playback of a discussion session.

(1902, 1906) which denotes the development of knowledge-gathering processes. Baldwin influenced the well-known and complicated psychophilosophical genetic epistemology of Jean Piaget (1952a,b). The philosophic nature of Piaget's writings, based upon systematic naturalistic inquiry, sometimes seems difficult for the more operationally oriented science of psychology.[2] Sutton-Smith (1966), for example, questioned some fundamental aspects of play and imitation in his interpretation of Piaget. Piaget (1966) responded by implying that Sutton-Smith did not understand him and that he did not recognize the conceptualizations attributed to him by Sutton-Smith. Furth (1968) also attempted to explain the nature of representation and interiorization in Piagetian theory, as have Berlyne (1965), Brown (1965), Flavell (1963), Fowler (1966), Hunt (1961), Kohlberg (1966), Smedslund (1961), and Wohwill (1966), to mention only a few of the scholars currently stimulated by the theory and research flowing from Geneva, who are generally associated with "cognition" or "cognitive development" in children and infants.

THEORY AND VERIFICATION

An era of great activity is beginning in developmental psychology, both on the theoretical and applied levels. Cohen and Nagel (1934) logically supported Pratt's (1948) contention that psychology, like all science, depends on both theory and verification. For many years, psychology has emphasized verification and generally neglected the formulation and evolution of theory. It may be that the passing of Kenneth W. Spence, perhaps one of the greatest of all empirical thinkers, will mark the beginning of a new emphasis, characterized by an increasing acceptance of Piaget's nonexperimental approach to developmental psychology.

In the United States, Jerome S. Bruner, from the Center for Cognitive Studies at Harvard, has championed this emphasis. Bruner has functioned more within the traditional scientific model familiar to psychologists; Looft and Bartz (1969) de-

---

[2] Researchers within the Piagetian frame of reference have found it difficult to establish reliable and valid ordinal scales of development. This problem was discussed at an Invitational Conference on Ordinal Scales of Development, sponsored by the California Test Bureau, February, 1969, where Prof. Peter M. Bentler (University of California, Los Angeles) presented two important papers: "An implicit metric for ordinal scales: Implications for assessment of cognitive growth," and "Monotonicity analysis: An alternative to linear factor and test analysis."

scribed him as an "information processing" theorist. His interest
in the problems of education may well be an expression of the
pragmatic tradition of American psychology responding to social
pressures and needs.

Bruner and Piaget differ most visibly in their interest in pedagogy. Piaget (Ripple and Rockcastle, 1964) offered little advice
to educators, while Bruner (1964a,b,c) envisioned the possibilities of psychology when it comes to terms with and assumes its
rightful role in education. Bruner said that psychology tends to
be somewhat provincial; he believes it should provide a theory
and a model for testing hypotheses regarding instruction.

Intelligent Behavior and Education

Children's needs in relation to the development of intelligent
behavior are monumental (Hickerson, 1966; Jensen, 1969; Rosenthal and Jacobson, 1968a,b). This is particularly true for "disadvantaged" children. Underlying problems perpetuate themselves in large cities and trouble spots such as Appalachia. National efforts such as Head Start and local attempts such as New
York City's "Higher Horizons" and "More Effective Schools"
have tried to compensate for the debilitating inadequacies with
which these children begin school.

THE VISIBLE BEHAVIORAL PROBLEMS

The inadequacy of poor environments is relatively well understood. For example, in San Antonio, Texas, at the initiation
of a bilingual (Spanish and English) reading project, a large
majority of the Latin-American children scored so low on a
reading readiness test that they were considered "unmeasurable."
In El Paso, Texas, a survey revealed that many Latin-American
preschoolers had never been beyond Piasano Street, the un-
official boundary of that city's Mexican-American ghetto. Conditions in other states are similar.[3]

[3] See the discussion of "Equal Educational Opportunity" in a special
issue of the *Harvard Educational Review* (1968), led by H. Howe, J. Cole-
man, and D. Moynihan. *TRANS-action* (February, 1969) devoted an entire
issue to "The American Underclass: Red, White, and Black," in which urban
poverty (Lipsky, Suttles, and Kochman), Mexican-American problems
(Love), the repression of the Oklahoma Cherokee (Wahrhaftig & Thomas),
and indiscriminate wastes of human labor among migrants in New York
State (Friedland) were discussed.

Illustrative data are available from a series of focused interviews (McGuire and White, 1955) conducted in lower-class homes in New York City and in Austin, Texas, which revealed criteria of social class as "markers" of probable variations in life experiences among children of families of different life-styles. After suitable orientation, the interviews were required in undergraduate educational psychology and child development courses, as prerequisites to teacher certification. Prospective educators discovered that among lower-class people, education was often only a means to a "better job." There was little or no expressed desire for a child to reach his potential, nor any expression of what Berlyne (1965) identified as epistemic curiosity. A few of the student-teacher interviewers wrote that:

> He told me that I would be a good teacher if I knew all the cuss words the kids did and would use them while I was teaching. He said that he had a son that was really dumb, but he made up for it with his cussing powers. . . . He said that the teachers had been giving his boy the worst grades in everything, and he decided that school really wasn't going to help him any.

> I asked her why she thought that her children had not graduated. She said that perhaps the fact that she could not buy them the things they wanted and needed was the main reason. . . . Before I left I asked her to tell me one thing she would want me to know since I was going to be a teacher. She thought a minute, and then she told me to spank the children at school. She told me that she thought teachers were afraid to spank them.

> Education to these families is not just matter-of-fact; it is a struggle and something distant. . . . These parents seem aware that their children will have to have some sort of an education to get a decent job.

As the last sample illustrates, members of the culture of poverty have had some contact with education. Someone, somewhere, failed miserably. Our science shares in the responsibility for developing descriptive and explanatory theories applicable to children's educational encounters. Unless such theories are evolved and published, there seems little hope of breaking the cycle in which these people are trapped.

Psychological Purposes in Education

The principles of operation in education should serve at least two fundamental purposes: to help learners control their environment and to help teachers increase their professional efficiency. Effective teachers will need a greater understanding of behavior, both cognitive and emotional, in order to identify the modifications (behavioral objectives) which are the goals of education. When educators accept behavior as their appropriate domain, they will necessarily restructure their "professional self-concept" and recognize education more as a behavioral science than as a practical "art."

As behavioral science, education and psychology share subject matter and methods. Psychology can envision the possible; education formulates hypotheses and applies those which have been tested and found to be practical. When this occurs, there will be considerably more emphasis on science and scientific methodology in education and less on its artistic dimensions which, though often very real, tend to be highly individual and resistant to transmission.

CURRICULUM AND INTELLIGENT BEHAVIOR

As tools, frames of reference, or extensions of the teacher which allow students to control a specific area (e.g., language or math), curricular experiences constitute the structure of the educational encounter. In this inevitable encounter,[4] the psychological objective of control is the same, though specific areas demand relevant methodologies. This objective is intelligent behavior, "an adaptive act controlled by a central process which responds according to the learned meaning attached to environmental symbols or objects" (Rowland, 1968, 1969). Acceptance of a common psychological objective for curricular experiences allows the psychologist or teacher to be less concerned with subject matter and, consequently, more concerned with developing intelligent behavior.

Theoretical principles regarding the development of intelligent behavior have been arrived at by means of a convergent analysis of available theories. The behavioral processes in the development of intelligent behavior rest on the assumption of a biological capability and necessity for the learning organism to interact with one or more of the genetic, internal, or external

[4] Every person faces a series of inevitable human encounters posed by his biological nature and the institutions necessary to regulate the behavior of man; for example: the transition from infancy to a nearly autonomous child, puberty and the change in body image, sex-typed expectations in the age-mate and adult societies, marriage and the "prime human encounter" which creates and "brings up" each individual replacement, then later maturity and death.

environments. The figure is a heuristic representation of the psychological processes common to *all* learning situations. These processes form an invariant sequence which begins with interaction from which the organism gains experience and ends in intelligent behavior. The acquisition of experience develops the central processes of control.

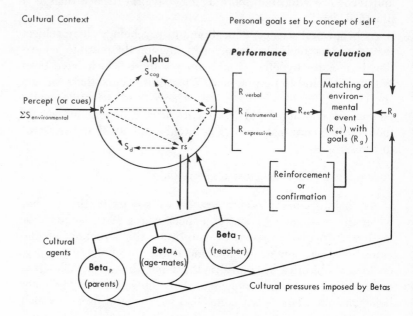

The dyadic model illustrates the psychological processes of reciprocal stimulation involved in human learning.

The Cultural Context of Learning

Human learning occurs in cultural contexts which involve the learner (Alpha) and certain cultural agents (Betas), such as parents, peers, and teachers. This model represents the dynamics of the educational encounter, wherein cognitive re-structuring ($S_{cog} \rightarrow S'_{cog}$), attitude change ($rs \rightarrow rs'$), and behavior modification ($R_{ee} \rightarrow R'_{ee}$) are instances of learning.

Each person brings to the learning situation (such as a classroom) an idiosyncratic level of perceptual readiness ($S \rightarrow R'$) as well as the capacity to respond (S_d), which may be conceived of being composed of "needs," "epistemic curiosity," or "emotional arousal" in the reticular activation system (RAS). The learner also brings to the learning situation the residuals of prior experience, the capacity to respond expressively, instrumentally, and verbally to mediated cues $(S' \rightarrow R)$, and habit patterns $(S-(\rightarrow R': :S'-) \rightarrow R_2)$ which reduce the amount of necessary mediation $(R'::S)$.

Each learner has acquired and is cued by expectancies $(rs_p'$ rs_a' $rs_t)$ of the supportive or nonsupportive behavior of the cultural agents, which may be understood to be a system of reward control by means of which a culture or subculture shapes behavior. These expectancies sometimes are studied as attitudes. Briefly, then, learning involves changes in perception, drive structure, cognitive structures, expectancies, and habit patterns, and is marked by observable modifications in performance. Goals (R_g) are set by the learner and/or by the pressures of cultural agents. When expectations are confirmed by the matching of a response or performance with goals, there is reinforcement of learning; when contradicted, there is either a redirection of behavior, or forms of inhibition and extinction.

The model is intentionally structured with "empty symbols" so that behavioral scientists of different orientations are able to impose their own conceptions. The model is not "closed" or complete, nor is the theory which accompanies it finished. The basic purpose of this model and theory is to allow each behavioral scientist to gain some insights into the development of human behavior patterns. Both model and theory are open to refinement. A student of Krasner and Ullman's *Research in Behavior Modification* (1967), for example, should be able to use it.

The Principles of Intelligent Behavior

Within the proposed theoretical context and its dyadic model, the following sequential principles are relevant: interaction, experience, and the central processes of control. These principles are sequential prerequisites to intelligent behavior which, in the school, would be the expression of some predetermined behavior in an observable manner.

Interaction consists of the reciprocal processes of organism-environment(s) feedback by means of which the organism directs, defines, redirects or redefines the enviroment(s) and its own behaviors according to a learned frame of reference. Each interactive experience serves additionally to differentiate the organism from the environment by means of change, both as transformations within the organism and as alterations of the environment(s) with which the organism interacts. Each interactive episode also provides the learner with experience or experiential information by means of which the central processes are ultimately developed and control of one's own behavior is extended.

Experience is the relationship of familiarity with the organism and its three interacting environments: biological heredity, one's self, and a social heritage (represented by cultural agents).

The central processes of control result from the organization and integration of informational feedback so that it can be retrieved upon organismic or environmental demand. For example, the "empty symbols" depicted within Alpha could be regarded as a reverberatory circuit involving cue-producing responses (R'), an arousal system (S_d), expectancies or attitudes (rs), cognitive schema (S_{cog}), and foresight or "feedforward" (S'). Self-instigation of intelligent behavior could be initiated at almost any point in the dynamic organization.

The Continuity-Discontinuity Principle

The invariant sequence which begins with interaction and culminates in intelligent behavior operates according to the principle of feedback (see Weiner, 1949). In this conceptualization, which except for the objective of equilibration [5] is quite compatible with Piaget's thinking, the learner initiates a primary behavior. Upon receiving negative or nonconfirming feedback, he initiates a secondary or accommodative behavior. Positive or confirming feedback results in no change and the continuance of assimilatory behavior. With these notions, education may be understood as the controlled introduction of discontinuity into the environmental interactions of learners.

[5] Altering the objective of equilibration would avoid the logical embarrassments of traditional homeostatic models. The refined objective is that the organism actively seeks stimulation or discontinuity and would therefore not find pleasure in a state of balance or inertia (i.e., achieved equilibrium). Instead, within this context, the learning and behaving organism remains in a constant state of dynamic and reciprocal interaction.

Summary

The psychologist is in an enviable position; he has the responsibility for producing theories which will permit the formulation and testing of hypotheses regarding the efficiency of instructional methods. The purpose of proposing a theory is to encourage colleagues to gather and verify facts and to initiate research. The theory of intelligent behavior provides the educator and psychologist with a set of sequential principles underlying the dynamics of the development of intelligent behavior. Applied research should begin with the most fundamental principle: interaction. If this variable can be controlled scientifically, certain observational behaviors should be forthcoming, further establishing and clarifying the role of the psychologist in the classroom.

Bibliography

ALLPORT, G. G. 1967. *Pattern and growth in personality.* New York: Holt, Rinehart & Winston, Inc. (Revision of *Personality: a psychological interpretation,* 1937.)

ALLPORT, G. W. 1966. Traits revisited. *American Psychologist,* **21**: 1–10.

ANDERSON, R. C. 1967. Educational psychology, in P. R. Farnsworth, O. McNemar, and Q. McNemar, eds., *Annual review of psychology.* Palo Alto: Annual Reviews, p. 158.

———, and D. P. AUSUBEL, eds. 1966. *Readings in the psychology of cognition.* New York: Holt, Rinehart & Winston, Inc.

ASHBY, W. R. 1960. *Design for a brain.* New York: John Wiley & Sons, Inc.

AUSUBEL, D. P. 1960a. Acculturative stress in modern Maori adolescents. *Child Development,* **31**: 617–31.

———. 1960b. The use of advance organizers in the learning of meaningful verbal material. *Journal of Educational Psychology,* **51**: 267–72.

———. 1963. A teaching strategy for culturally deprived pupils: Cognitive and motivational considerations. *School Review,* **71**: 454–63.

———. 1966. Review of J. Bruner's *Toward a theory of instruction. Harvard Educational Review,* **36**: 337–40.

BALDWIN, A. L. 1968. Role of an "ability" construct in a theory of behavior, in D. C. McClelland, A. L. Baldwin, U. Bronfenbrenner, and F. L. Strodebeck, *Talent and society: new perspectives in the identification of talent.* Princeton, N.J.: D. Van Nostrand Co., Inc., pp. 195–233.

BALDWIN, J. M. 1902. *Development and evolution.* New York: The Macmillan Company.

———. 1906. *Mental development.* New York: The Macmillan Company.

BANDURA, A., and F. J. McDONALD. 1963. The influence of social reinforcement and the behavior of models in shaping children's moral judgments. *Journal of Abnormal and Social Psychology,* **67**: 274–81.

BAYLEY, N. 1949. Consistency and variability in the growth from birth to 18 years. *Journal of Genetic Psychology,* **75**: 165–96.

BEACH, F. A. 1945. Current concepts of play in animals. *American Naturalist,* **79**: 523–41.

BEADLE, G., and M. BEADLE. 1966. *The language of life: an introduction to the science of genetics.* New York: Doubleday & Company, Inc.

BERGMANN, G. 1953. Theoretical psychology. *Annual Review of Psychology,* **4**: 435–38.

BERLYNE, D. E. 1954a. An experimental study of human curiosity. *British Journal of Psychology,* **45**: 256–65.

———. 1954b. A theory of human curiosity. *British Journal of Psychology.* **45**: 180–91.

———. 1957. Conflict and information theory variables as determinants of human perceptual curiosity. *Journal of Experimental Psychology.* **53**: 399–404. Reprinted in Harry Fowler, ed., *Curiosity and exploratory behavior.* New York: The Macmillan Company, 1965, pp. 191–99.

———. 1962. Uncertainty and epistemic curiosity. *British Journal of Psychology,* **53**: 27–34.

———. 1963. Motivational problems raised by exploratory and epistemic behavior, in S. Koch, ed., *Psychology: A study of a science. The process areas, the person, and some applied fields: their place in psychology and in science.* New York: McGraw-Hill Book Company, **V**: 284–364.

———. 1965. *Structure and direction in thinking.* New York: John Wiley & Sons, Inc.

———. 1966. Conflict and arousal. *Scientific American,* pp. 82–87.

BERNSTEIN, N. 1967. *The coordination and regulation of movement.* New York: Pergamon Press.

BEXTON, W. H., W. HERON, and T. H. SCOTT. 1954. Effects of decreased variation in the sensory environment. *Canadian Journal of Psychology,* 70–76.

BINET, A. 1909. *Les idées modernes sur les enfants.* Paris: Ernest Flamarion, in G. D. Stoddard. The IQ: Its ups and downs. *Educational Record,* 1939, **20**: 44–57.

BIRCH, H. G. 1956. Sources of order in maternal behavior of animals. *American Journal of Orthopsychiatry,* **26**: 179–84.

BLUMER, H. 1962. Society as symbolic interaction, in A. M. Rose, ed., *Human behavior and social processes.* Boston: Houghton Mifflin Company, pp. 179–92. Republished in Manis, G. N., and B. N. Meltzer, eds., *Symbolic Interaction.* Boston: Allyn & Bacon, Inc., 1967, pp. 139–48.

BORING, E. G. 1957. *A history of experimental psychology.* New York: Appleton-Century-Crofts.

BRONFENBRENNER, U. 1962. Soviet studies of personality development and social psychology, in R. A. Bauer, ed., *Some views on Soviet psychology.* Washington, D.C.: American Psychological Association, pp. 63–86.

BROWN, R. 1965. *Social psychology.* New York: The Free Press.

BROZEK, J. 1964. Recent developments in Soviet psychology. *Annual Review of Psychology,* **15**: 493–594.

BRUNER, J. S. 1956. A cognitive theory of personality. *Contemporary Psychology,* **1**: 355–56.

———. 1960. *The process of education,* New York: Vintage.

———. 1961a. The act of discovery. *Harvard Educational Review,* **31**: 21–32.

———. 1961b. The growth of mind. *American Psychologist,* **1**: 1–15.

———. 1964a. The course of cognitive growth. *American Psychologist,* **19**: 1–15.

———. 1964b. Education as social invention. Invited address to the 9th Inter-American Congress of Psychology, Miami Beach.

———. 1964c. The ferment in American education, introduction to De Grazie, A., and D. A. Sohn, eds., *Revolution in teaching: new theory, technology, and curricula.* New York: Bantam Books, Inc.

———. 1965. The growth of mind. *American Psychologist,* **20**: pp. 1007–17.

———. 1966. *Toward a theory of instruction.* Cambridge, Mass.: Belknap.

———. 1967. Origins of mind in infancy. Address to Division 8 at the 75th Anniversary of the American Psychological Association, Washington, D.C.

———. 1968a. Culture, politics, and pedagogy. *Saturday Review* (May 18), 69–72f.

———. 1968b. The growth and structure of skill. Ciba Conference, London.

―――. 1968c. *Processes of cognitive growth: Infancy*. Heinz Werner Lecture Series. Worcester, Mass.: Clark University Press, III.

―――. 1969a. The psychobiology of pedagogy. Sigma Xi Lecture at Rockefeller University, New York.

―――. 1969b. Origins of problem solving strategies in skill acquisition. Presented at the XIX International Congress of Psychology, London.

―――. 1969c. Up from helplessness. *Psychology Today*, 8: 30f.

―――. 1969d. Eye, hand and mind, in D. Elkind and J. H. Flavell, eds. *Studies in cognitive development: Essays in honor of Jean Piaget*. New York: Oxford University Press, Inc., pp. 223–35.

―――, and B. M. BRUNER. 1968. On voluntary action and its hierarchical structure. Presented at the Symposium on New Perspectives in the Sciences of Man, Alpbach, Austria.

BRUNER, J. S., J. J. GOODNOW, and G. A. AUSTIN. 1959. *A study of thinking*. New York: Science Editions.

BRUNER, J. S., K. KAYE, and K. LYONS. 1970. The growth of human manual intelligence: III. The development of detour tracking, in G. T. Rowland and J. Anglin, eds., *Bruner: Essays on cognition and education*. New York: W. W. Norton & Company, Inc.

BRUNER, J. S., and H. J. KENNEY. 1965. Representation and mathematics learning. *Monograph of the Society for Research in Child Development*, 30: (Serial No. 99), 50–59.

BRUNER, J. S., R. R. OLVER, P. M. GREENFIELD, J. R. HORNSBY, H. J. KENNY, M. MACCOBY, N. MODIANO, F. A. MOSHER, D. R. OLSON, M. C. POTTER, L. C. REICH, and A. McK. SONSTROEM. 1966. *Studies in cognitive growth*. New York: John Wiley & Sons, Inc.

BRUNER, J. S., J. MATTER, and M. L. PAPANEK. 1955. Breadth of learning as a function of drive level and mechanization. *Psychological Review*, 62: 1–10.

BRUNER, J. S., J. SIMENSON, and K. LYONS. 1970. The growth of human manual intelligence: I. Taking possession of objects, in G. T. Rowland and J. Anglin, eds., *Bruner: Essays on cognition and education*. New York: W. W. Norton & Company, Inc.

BRUNER, J. S., and D. WATKINS. 1970. The growth of human manual intelligence. Acquisition of complementary two-handedness, in G. T. Rowland and J. Anglin, eds., *Bruner: Essays on cognition and education*. New York: W. W. Norton & Company, Inc.

BURT, C. 1940. *The factors of the mind*. London: University of London Press.

―――. 1949. The structure of the mind: A review of the results of factors analysis. *British Journal of Educational Psychology*, 19: 100–11, 176–99.

146 BIBLIOGRAPHY

CANNON, W. B. 1939. *The wisdom of the body.* New York: W. W. Norton & Company, Inc.

CASSIRER, E. 1961. *Essay on man.* New York: Dover.

————. 1968. *The myth of the state.* New Haven: Yale University Press.

CLARK, R. A., and C. McGUIRE. 1952. Sociographic analysis of sociometric valuations. *Child Development,* 23: 129–40.

COFER, C. N., and M. H. APPLEY. 1964. *Motivation: theory and research.* New York: John Wiley & Sons, Inc.

COHEN, M. R., and E. NAGEL. 1934. An introduction to logic and scientific method. New York: Harcourt, Brace & World, Inc.

COHEN, SHIRLEY. 1966. The problem with Piaget's child. *Teachers College Record,* 68(3): 211–18.

COLBY, K. M. 1967. Computer simulation of change in personal belief systems. *Behavioral Science,* 12: 248–53.

COLE, M., and I. MALTZMAN, eds. 1969. *A handbook of contemporary Soviet psychology.* New York: Basic Books, Inc., Publishers.

CONANT, J. B. 1959. *The American high school today.* New York: McGraw-Hill Book Company; Signet Book, 1964 (Paperback).

CRONBACK, L. J. 1967. To the Editor. *Harvard Educational Review,* 37(3): 468.

CRUZE, W. W. 1935. Maturation and learning in chicks. *Journal of Comparative Psychology,* 19: 371–409.

DARWIN, D. 1872. *The expressions of the emotions in man and animals.* London: John Murray (Publishers) Ltd. (New York: Appleton, 1873).

DENNIS, W., and P. NAJARIAN. 1957. Infant development under environmental handicap. *Psychological Monographs,* 71: Whole No. 46.

DOLLARD, J., and N. E. MILLER. 1950. *Personality and psychotherapy: An analysis in terms of learning, thinking and culture.* New York: McGraw-Hill Book Company.

DONALDSON, M. 1963. *A study of children's thinking.* London: Tavistock Publications.

ELKIND, D. 1969. Piagetian and psychometric conceptions of intelligence. *Harvard Educational Review,* 39: 319–37.

ENGLISH, H. B., and A. C. ENGLISH. 1958. *A comprehensive dictionary of psychological and psychoanalytical terms: A guide to usage.* New York: Longmans, Green.

ERIKSON, E. H. 1950. *Childhood and society,* 2nd ed. New York: W. W. Norton & Company, Inc.

FERGUSON, G. A. 1954. On learning and human ability. *Canadian Journal of Psychology,* 8: 95–112.

————. 1956. On transfer and the abilities of man. *Canadian Journal of Psychology,* **10:** 121–31.

————. 1965. Human abilities. *Annual Review of Psychology,* **16:** 39–62.

FESTINGER, L. 1957. *A theory of cognitive dissonance.* New York: Harper & Row, Publishers.

FLAVELL, J. H. The developmental psychology of Jean Piaget. New York: Van Nostrand, 1963.

FLEISHMAN, E. A., and W. E. HEMPEL. 1954. Changes in factor structure of a complex psychomotor test as a function of practice, *Psychometrika,* **19:** 239–52.

FLETCHER, J. M. 1938. The wisdom of the mind. *Sigma Xi Quarterly,* **26:** 6–16.

————. 1942. Homeostasis as an explanatory principle in psychology. *Psychological Review,* **49:** 80–87.

FOA, U. G. 1965. New developments in facet design and analysis. *Psychological Review,* **72:** 262–72.

FOWLER, W. 1966. Dimensions and directions in the development of affectocognitive systems. *Human Development,* **9:** 18–29.

FRANKFORT, H. 1949. *Before philosophy.* London: Pelican.

FREEMAN, G. L. 1948. *The energetics of human behavior.* Ithaca, N.Y.: Cornell University Press.

FRENCH, J. D. 1957. The reticular formation. *Scientific American,* pp. 55–56.

————, R. HERNANDEZ-PEON, and R. B. LIVINGSTON. 1955. Projections from cortex to cephalic brain stem (reticular formation) in monkeys. *Journal of Neurophysiology,* **18:** 44–45.

FREUD, S. 1926. *Hummung, Symptom, and Angst.* (Translated as *The problem of anxiety* by H. A. Barker. New York: W. W. Norton & Company, Inc., 1936.)

FROST, J. L., and G. T. ROWLAND. 1968a. Cognitive development and literacy in disadvantaged children: A structure-process approach. In J. L. Frost, ed., *Early childhood education rediscovered.* New York: Holt, Rinehart & Winston, Inc.

————. 1968b. *Elaborative language series II.* Austin, Texas: Educational Media Laboratories.

————. 1969. *Curricula for the seventies.* Boston: Houghton Mifflin Company.

FURTH, H. G. 1968. Piaget's theory of knowledge: The nature of representation and interiorization. *Psychological Review,* **75:** 143–54.

GAGNÉ, R. M. 1962. The acquisition of knowledge. *Psychological Review,* **69:** 355–65.

————. 1965a. *The conditions of learning*. New York: Holt, Rinehart & Winston, Inc.

————. 1965b. The learning of concepts. *School Review*, **73**: 187–96.

————, and N. E. PARADISE. 1961. Abilities and learning sets in knowledge acquisition. *Psychological Monographs*, **75**: Whole No. 518.

GARDNER, J. W. 1961. *Excellence: Can we be equal and excellent too?* New York: Harper Colophon.

GARRETT, H. E. *Children: black & white*. Undated Monograph. Kilmarnock, Virginia: The Patrick Henry Press, Inc.

GAZZANIGA, M. 1967. The split brain in man. *Scientific American*, **217**.

GESELL, A. 1943. *The embryology of behavior: The beginning of the human mind*. New York: Harper & Row, Publishers.

GLASS, D. C. 1967. Genetics and social behavior. *Items*, **21**: 1–5.

GOODNOW, J. J. 1962. A test of milieu effects with some of Piaget's tasks. *Psychological Monographs*, **76**: Whole No. 555.

GREENFIELD, P. 1969a. Goal as Environmental Variable in the Development of Intelligence. Paper presented to the Conference on Contributions to Intelligence. Urbana: The University of Illinois. November 15, 1969.

————. 1969b. Personal communication.

GRIER, J. B., S. A. COUNTER, and W. M. SHEARER. 1967. Prenatal auditory imprinting in chickens. *Science*, **155**: 1692–93.

GUILFORD, J. P. 1956. The structure of the intellect. *Psychological Bulletin*, **53**: 267–93.

————. 1959a. *Personality*. New York: McGraw-Hill Book Company.

————. 1959b. Three faces of intellect. *American Psychologist*, **14**: 469–79.

————. 1961. Factorial angles to psychology. *Psychological Review*, **68**: 1–20.

GUTTMAN, L. 1958. What lies ahead for factor analysis? *Educational and Psychological Measurement*, **18**: 497–515.

HALL, C. S. 1954. *A primer of Freudian psychology*. New York: Mentor.

HARLOW, H. F. 1949. The formation of learning sets. *Psychological Review*, **56**: 51–65.

————. 1950. Learning and satiation of response in intrinsically motivated complex puzzle performance by monkeys. *Journal of Comparative Physiological Psychology*, **43**: 289–94.

————. 1951a. Primate learning, in C. P. Stone, ed., *Comparative psychology*. New York: Prentice-Hall, Inc.

————. 1951b. Thinking, in H. Helson, ed., *Theoretical foundations of psychology*. Princeton, N.J.: D. Van Nostrand Co., Inc.

————. 1953. Mice, monkeys, men and motives. *Psychological Review*, **60**: 23–32.

————, and M. K. HARLOW. 1949. Learning to think. *Scientific American*, **181**: 36–39.

HAVIGHURST, R. J., and H. TABA. 1949. *Adolescent character and personality*. New York: John Wiley & Sons, Inc.

HEBB, D. O. 1942. The effect of early and late brain injury upon tests scores, and the nature of normal adult intelligence. *Proceedings of the American Philosophical Society*, **85**: 275–92.

————. 1946. On the nature of fear. *Psychological Review*, **53**: 259–76.

————. 1949. *The organization of behavior*. New York: John Wiley & Sons, Inc.

————. 1955. Drives and the C.N.S. (Conceptual Nervous System). *Psychological Review*, **62**: 243–54.

————. 1958. *A textbook of psychology*. Philadelphia: W. B. Saunders Company.

————. 1960. The American revolution. *American Psychologist*, **9**: 107–16.

————, and W. R. THOMPSON. 1954. The social significance of animal studies, in G. Lindgey, ed., *Handbook of Social Psychology*. Reading, Mass.: Addison-Wesley Publishing Co., Inc.

HEIDER, F. 1946. Attitudes and cognitive organization. *Journal of Psychology*, **21**: 107–12.

HELSON, H. 1947. Adaptation-level as frames of reference for prediction of psycholophysical data. *American Journal of Psychology*, **60**: 1–29.

————. 1948. Adaptation-level as a basis for a quantitative theory of frame of reference. *Psychological Review*, **55**: 297–313.

HESS, H. H. 1962. Ethology: An approach toward the complete analysis of behavior, in *New directon in psychology 1*. New York: Holt, Rinehart & Winston, Inc., pp. 155–266.

HICKERSON, N. 1966. *Education for alienation*. Englewood Cliffs, N.J.: Prentice-Hall, Inc.

HILGARD, E. R. 1966. *Theories of learning*. New York: John Wiley & Sons, Inc.

————, and G. H. BOWER. 1943. *Theories of learning*, 3rd ed. New York: Appleton-Century-Crofts.

HINDSMAN, E., and R. L. DUKE. 1948. Development and utilization of talent. *Journal of Experimental Education*, **4**: 309–24.

HIRSCH, J. 1967. Behavior genetics, or "experimental" analyses: the challenge of science versus the use of technology. *American Psychologist,* **22**: 118–30.

HOLLINGSHEAD, A. B. 1949. *Elmtown's youth: the impact of social classes on adolescents.* New York: John Wiley & Sons, Inc.

HONZIK, M. P., J. W. MACFARLANE, and L. ALLEN. 1948. The stability of mental test performance between 2 and 18 years. *Journal of Experimental Education,* **17**: 309–24.

HUDSON, W. 1967. The study of the problem of pictorial perception among unacculturated groups. *International Journal of Psychology,* **2**: 89–102.

HULL, C. L. 1931. Goal attraction and directing ideas conceived as habit phenomena. *Psychological Review,* **38**: 487–506.

————. 1934. *Principles of behavior.* New York: Appleton-Crofts.

————. 1952. *A behavior system: an introduction to behavior theory concerning the individual organism.* New Haven: Yale University Press.

HUMPHREYS, L. G. 1962. The organization of human abilities. *American Psychologist,* **17**: 475–83.

HUNT, J. McV. 1941. The effects of infantile feeding-frustration upon adult hoarding in the albino rat. *Journal of Abnormal Social Psychology,* **36**: 338–60.

————. 1946. Experimental psychoanalysis, in P. L. Harriman, ed., *Encyclopedia of psychology.* New York: Philosophical Library.

————. 1960. Experience and the development of motivation: Some reinterpretations. *Child Development,* **31**: 489–504.

————. 1961. *Intelligence and experience.* New York: The Ronald Press Company.

————. 1964. The psychological basis for using pre-school enrichment as an antidote for cultural deprivation. *Merrill-Palmer Quarterly of Behavior and Development,* **10**. In F. M. Hechinger, ed., *Pre-school education today.* New York: Doubleday & Company, Inc., 1966.

————. 1965. Traditional personality theory in the light of recent evidence. *American Scientist,* **53**: 80–96.

INHELDER, B., and J. PIAGET. 1958. *The growth of logical thinking.* New York: Basic Books, Inc., Publishers.

JENSEN, A. R. 1969. How much can we boost IQ and scholastic achievement? *Harvard Educational Review,* **39**: 1–123.

JOHNSON, E. E. 1953. The role of motivational strength in latent learning. *Journal of Comparative and Physiological Psychology,* **45**: 526–30.

KELLY, F. J., and J. J. CODY. 1969. *Educational Psychology: a behavioral approach.* Columbus, Ohio: Charles E. Merrill.

KELLY, G. A. 1955. *The psychology of personal constructs.* New York: W. W. Norton & Company, Inc.

KENDLER, T. S. 1963. Development of mediating responses in children. In *Monograph of the Society for Research in Child Development,* No. 28: Whole No. 86, 33–48. Reprinted in R. C. Anderson and D. P. Ausubel, eds., *Readings in the psychology of cognition.* New York: Holt, Rinehart & Winston, Inc., pp. 501–20.

KEPPEL, F. 1966. *The necessary revolution in American education.* New York: Harper & Row, Publishers.

KERSH, B. Y. 1958. The adequacy of "meaning" as an explanation for the superiority of learning by independent discovery. *Journal of Educational Psychology,* **49**: 282–92.

————. 1962. The motivational effect of learning by directed discovery. *Journal of Educational Psychology,* **53**: 65–71.

KLUCKHOHN, C., and H. A. MURRAY, eds. 1953. *Personality in nature, society, and culture,* 2nd ed. New York: Alfred A. Knopf, Inc.

KOHLBERG, L. 1966. Cognitive stages and preschool education. *Human Development,* **9**: 5–17.

KOMISAR, T. P. 1966. The paradox of equality in schooling. *Teachers College Record,* **68**: 251–54.

KRASNER, L., and L. P. ULLMAN. 1967. *Research in behavior modification.* New York: Holt, Rinehart, & Winston, Inc.

KRECHEVSKY, I. 1932. "Hypotheses" versus "chance" in the pre-solution period in sensory discrimination-learning. *University of California Publications in Psychology,* **6**: 27–44.

LERNER, D., and H. D. LASSWELL, eds. 1951. *The policy sciences: recent developments in scope and method.* Stanford, Calif.: Stanford University Press.

LEWIN, K. 1935. *A dynamic theory of personality.* New York: McGraw-Hill Book Company.

————. 1946. Behavior and development as a function of the total situation, in L. Carmichael, ed., *Manual of child psychology.* New York: John Wiley & Sons, Inc., pp. 791–845.

————. 1951. *Field theory in social science.* New York: Harper & Row, Publishers.

LEWIS, O. 1961. *The children of Sanchez.* New York: Random House, Inc.

————. 1966. The culture of poverty. *Scientific American,* 1966, **215**, 19–25.

LINDSLEY, D. B. 1950. Emotions and the electroencephalogram. In M. L. Reymert, ed., *Feelings and emotions.* New York: McGraw-Hill Book Company, pp. 238–46.

————. 1951. Emotion. In S. S. Stevens, ed., *Handbook of experimental psychology*. New York: John Wiley & Sons, Inc., pp. 473–561.

————. 1957. Psychophysiology and motivation, in M. R. Jones, ed., *Nebraska symposium on motivation*. Lincoln: University of Nebraska Press, pp. 44–105.

————, J. BOWDEN, and H. W. MAGOUN. 1949. Effect upon the EEG of acute injury to the brain stem activating system. *EEG and Clinical Neurophysiology*, **1**: 475–86.

LOOFT, W. R., and W. BARTZ. 1969. II. Animism reviewed. *Psychological Bulletin*, **71**: 1–19.

LURIA, A. R. 1959. The directive function of speech in development and dissolution. *Word*, **15**: 341–52. Reprinted in R. C. Anderson and D. P. Ausubel, eds., *Readings in the psychology of cognition*. New York: Holt, Rinehart & Winston, Inc. 1966.

MCBEE, G., and R. L. DUKE. 1960. Relationship between intelligence, scholastic motivation, and academic achievement. *Psychological Reports*, **6**: 3–8.

MCCLELLAND, D. C. 1951. *Personality*. New York: William Sloand (Dryden).

————. 1965. Toward a theory of motive acquisition. *American Psychologist*, **20**: 321–33.

————, A. L. BALDWIN, U. BRONFENBRENNER, and F. L. STRODE-BECK. 1958. *Talent and society: new perspective in the identification of talent*. Princeton, N.J.: D. Van Nostrand Co., Inc.

MCDONALD, F. J. 1959. *Educational psychology*. Belmont, Calif.: Wadsworth Publishing Co. Inc.

MCGUIRE, C. 1945. Education for the wider community. *The School* (Canada), **34**: 16–20.

————. 1949. Adolescent society and social mobility. Ph.D. Dissertation, University of Chicago.

————. 1956. The Textown study of adolescence. *Texas Journal of Science*, **8**: 264–74.

————. 1961. Sex role and community variability in test performances. *Journal of Educational Psychology*, **52**: 61–73.

————. 1965. Commentaries, in Mary Jane Aschner and C. E. Bish, eds., *Productive thinking in education, part II: motivation, personality, & productive thinking*. Washington, D.C.: National Education Association, pp. 180–90.

————. 1967. Behavioral science memorandum no. 13 of the Research and Development Center for Teacher Education at the University of Texas, Austin, mimeographed.

————. 1969. *The years of transformation*. Final report to the U.S.

Department of Health, Education, and Welfare, Office of Education, Bureau of Research, Project number 742. Washington, D.C.: Government Printing Office.

——, and R. A. Clark. 1952. Age-mate acceptance and peer status. *Child Development,* **23**: 141–54.

McGuire, C., and T. Rowland. *Behavioral science foundations of education: a handbook for learning to guided discovery.* New York: Holt, Rinehart & Winston, Inc. (in process).

McGuire, C., and G. W. White. 1955. The measurement of social status. Research paper in human development number 3 (revised). Department of Educational Psychology, University of Texas, Austin.

McNemar, Q. 1940. A critical examination of the University of Iowa studies of environmental influences upon the IQ. *Psychological Bulletin,* **37**: 63–92.

Maltzman, I. 1955. Thinking: from a behavioristic point of view. *Psychological Review,* **62**: 457–67.

Milerian, E. A. 1960. Psychological characteristics of the transfer of technical skills in older school children, trans. B. Simon and J. Simon, *Educational Psychology in the U.S.S.R.* Stanford, Calif.: Stanford University Press, 1963.

Miller, G. A. 1956. The magical number seven, plus or minus two: Some limits on our capacity for processing information. *Psychological Review,* **63**: 81–97. Reprinted in R. C. Anderson and D. P. Ausubel, eds., *Readings in the psychology of cognition.* New York: Holt, Rinehart & Winston, Inc., 1966, pp. 241–67.

——, E. H. Galanter, and H. H. Pribram. 1960. *Plans and the structure of behavior.* New York: Holt, Rinehart & Winston, Inc.

Miller, N. E. 1951. Comments on theoretical models illustrated by the development of a theory of conflict behavior. *Journal of Personality,* **20**: 82–100.

——, and J. Dollard. 1941. *Social learning and imitation.* New Haven: Yale University Press.

Morgan, C. 1909. *An introduction to comparative psychology.* London: Scott, 1909 (originally published in 1894).

——. 1965. *Physiological psychology,* 3rd ed. New York: McGraw-Hill Book Company.

Morozova, N. G. 1955. The psychological conditions for the arousal and modification of interest in older children in the process of reading popular scientific literature. *Izvestig. Akad. Ped. Nauk. RSFSR,* **73**: 100–149.

Moruzzi, G., and H. W. Magoun. 1949. Brain stem reticular formation and activation of the EEG. *Clinical Neurophysiology,* **1**: 455–73.

MULLER, H. J., C. C. LITTLE, and L. H. SNYDER. 1947. *Genetics, medicine and man*. Ithaca, N.Y.: Cornell University Press.

NEWELL, A., J. C. SHAW, and H. A. SIMON. 1958. Elements of a theory of human problem solving. *Psychological Review,* **65:** 151–66.

NEWMAN, H. H., F. N. FREEMAN, and K. J. HOLZINGER. 1937. *Twins: a study of heredity and environment*. Chicago: University of Chicago Press.

OLSON, D. R. 1966. On conceptual strategies, in J. S. Bruner, H. J. Kenny, M. Maccoby, N. Modiano, F. A. Mosher, D. R. Olson, M. C. Potter, L. C. Reich, and A. McK. Sonstroem. *Studies in cognitive growth*. New York: John Wiley & Sons, Inc.

OSGOOD, C. E. 1952. The nature and measurement of meaning. *Psychological Bulletin,* **49:** 192–237.

PARSONS, T., and E. A. SHILS, eds. 1951. *Toward a general theory of action*. Cambridge, Mass.: Harvard University Press.

PIAGET, J. 1923a. Une forme verbal de comparaison chez l'enfant (un cas de transition entre le jugement prédicatif et le jugement de relation). *Archives de Psychologie,* **18.**

———. 1923b. *The language and thought of the child*. (Trans. by M. Worden, New York: Harcourt, 1926.)

———. 1926. *The child's conception of the world*. (Trans. by J. Tomlinson and A. Tomlinson, New York: Harcourt, 1930.)

———. 1927. *The child's conception of physical causality*. (Trans. by M. W. Gabian, New York: Harcourt, 1930.)

———. 1932. *The moral judgment of the child*. (Trans. by M. W. Gabian, New York: Harcourt, 1932.)

———. 1936. *The origins of intelligence in children*. (Trans. by M. Cook, New York: International Universities Press, 1952.)

———. 1937. *The construction of reality in the child*. (Trans. by M. Cook, New York: Basic Books, Inc., Publishers, 1954.)

———. 1942. *Classes, relations et nombres. Essai sur les "groupements" de la logistique et la reversibilité de la pensée*. Paris: Brin.

———. 1945. *Play, dreams, and imitation in childhood*. (Trans. by C. Gattegno and F. M. Hodgson. New York: W. W. Norton & Company, Inc., 1951.)

———. 1947. *The psychology of intelligence*. (Trans. by M. Piercy and D. E. Berlyne. London: Routledge & Kegan Paul, Ltd., 1950.)

———. 1949. Le problème neurologique de l'interiorisation des actions en operations reversibles. *Archives de Psychologie* (Genève), **32:** 241–58.

————. 1952a. *Judgment and reasoning in the child.* New York: Humanities Press.

————. 1952b. Jean Piaget, in E. G. Boring, H. S. Lanfield, H. Werner, and R. M. Yerkes, eds., *A history of psychology in autobiography.* Worchester, Mass.: Clark University Press.

————. 1953. How children form mathematical concepts. *Scientific American.* Reprinted in P. H. Mussen, J. J. Conger, and J. Kegan, eds., *Readings in child development and personality.* New York: Harper & Row, Publishers, 1965.

————. 1961. A report on the conference on cognitive studies and curriculum development. Ithaca, N.Y.: Cornell University Press.

————. 1966. Response to Brian Sutton-Smith. *Psychological Review,* **73**: 111–12.

PRATT, G. C. 1948. *The logic of modern psychology.* New York: Macmillan.

REISS, B. F. 1950. The isolation of factors of learning and native behavior in field and laboratory studies. *Annual of the New York Academy of Science,* **51**: 1093–1103.

RICHTER, C. P. 1942. Total self-regulatory functions in animals and human beings. *Harvey Lectures,* **38**: 63–103.

RIPPLE, R. E., and V. N. ROCKCASTLE, eds. 1964. *Piaget Rediscovered.* A report on the Conference on Cognitive Studies and Curriculum Development. Ithaca: Cornell University School of Education, mimeographed.

ROGERS, C. R. 1951. *Client-centered therapy.* Boston: Houghton-Mifflin Company.

ROMANES, G. J. 1883. *Animal intelligence.* New York: Appleton.

————. 1884. *Mental evolution in animals.* New York: Appleton.

ROSENTHAL, R., and L. JACOBSON. 1968a. *Pygmalion in the classroom.* New York: Holt, Rinehart & Winston, Inc.

————. 1968b. Teacher expectations for the disadvantaged. *Scientific American,* **218**: 19–23.

ROSENZWEIG, M. R. 1966. Environmental complexity, cerebral change and behavior. *American psychologist,* **21**: 321–32.

ROUSSEAU, J. J. 1961. *The social contract.* New York: Pelican Books.

ROWLAND, G. T. 1967. Cognitive development in children: A model for intervention. Paper presented at the meeting of the Southwestern Psychological Association, Houston.

————. 1968. *Convergent foundations of a psychological theory of intelligent behavior.* Doctoral dissertation, University of Texas at Austin.

————. 1969. Cognitive development in children II: A structure-process approach. *Psychology in the schools,* **6**: 55–58.

————, and J. M. ANGLIN, eds. Forthcoming. *Bruner on man.* New York: W. W. Norton & Company, Inc.

ROWLAND, G. T., D. ELKIND, I. E. SIGEL, I. UZGIRIS, W. ROWHER, and P. GREENFIELD. 1969. Current thinking and research in the development of intelligent behavior. Symposium, Division 7, presented at the 77th annual convention of the American Psychological Association, Washington, D.C.

ROWLAND, G. T., and C. McGUIRE. 1968a. The development of intelligent behavior I: Jean Piaget. Psychology in the schools, **5** (1): 47–52.

————. 1968b. The development of intelligent behavior II: D. E. Berlyne. *Psychology in the schools,* **5** (2): 106–13.

————. 1968c. The development of intelligent behavior III: Robert W. White. *Psychology in the schools,* **5**: 230–39.

ROWLAND, G. T., and J. L. FROST. 1970. Human motivation: A structure-process interpretation. *Psychology in the schools.*

SAPIR, E. 1921. *Language.* New York: Harcourt, Brace & World.

SEARS, R. R. 1951. A theoretical framework for personality and social behavior. *American Psychologist,* **6**: 476–83.

————, J. W. M. WHITING, and I. L. CHILD. 1953. *Child training and personality.* New Haven: Yale University Press.

SEGALL, M. H., D. T. CAMPBELL, and M. J. HERSKOVOTZ. 1966. *Influence of culture on visual perception.* Indianapolis: The Bobbs-Merrill Co., Inc.

SIGEL, I. E. 1953. Developmental trends in the abstraction ability of children. *Child Development,* **24** (2): 131–44.

————. 1954. The dominance of meaning. *Journal of genetic psychology,* **85**: 201–8.

————. 1968a. The distancing hypothesis: a hypothesis crucial to the development of representational competence. Paper presented as part of the symposium, "Comparative Studies of Conceptual Functioning in Young Children," at the annual meeting of the American Psychological Association, San Francisco.

————. 1968b. The distancing hypothesis: A causal hypothesis for the acquisition of representational thought. Paper prepared for the symposium, "The Effects of Early Experience," The University of Miami, M. R. Jones, Chairman.

————, L. M. ANDERSON, and H. SHAPIRO. 1966. Categorization behavior of lower- and middle-class Negro preschool children: differences in dealing with representation of familiar objects. *Journal of Negro Education,* **35**: 218–29.

SIGEL, I. E., and B. MCBANE. 1967. Cognitive competence and level of symbolization among five-year-old children, in J. Hellmuth, ed., *The disadvantaged child,* Vol. 1. Seattle: Special Child Publications, 435–53.

SIGEL, I. E., and C. PERRY. 1968. Psycholinguistic diversity among "culturally deprived" children. *American Journal of Orthopsychiatry,* **38:** 122–26.

SKEELS, H. M., R. UPDEGRAF, B. L. WILLMAN, and H. M. WILLIAMS. 1938. A study of environmental stimulation: An orphanage preschool project. *University of Iowa Study of Child Welfare,* **15:** No. 4.

SKEELS, H. M. and H. B. DYE. 1939. A study of the effects of differential stimulation on mentally retarded children. *Proceedings of the American Association on Mental Deficiency,* **44:** 114–36.

SKINNER, B. F. 1950. Are theories of learning necessary? *Psychological Review,* **57:** 193–216.

SMEDSLUND, J. 1961. The acquisition of conservation of substance and weight in children III: Extinction of conservation of weight acquired "normally" and by means of empirical controls on a balanced scale. *Scandinavian Journal of Psychology,* **2:** 71–84.

SMILANSKY, S. 1968. *The effects of socio-dramatic play on disadvantaged pre-school children.* New York: John Wiley & Sons, Inc.

SPEARMAN, C. 1927. *The abilities of man.* New York: The Macmillan Company.

STAATS, A. W. 1961. Verbal habit-families, concepts, and the operant conditioning of word classes. *Psychological Review,* **68:** 190–204.

STEVENS, S. S., ed. 1951. *Handbook of experimental psychology.* New York: John Wiley & Sons, Inc.

STEVENSON, H. W., ed. 1966. Concept of development. *Monographs of the Society for Research in Child Development,* **31** (5) (Serial No. 107).

STOTT, L. H., and R. S. BALL. 1965. Infant and preschool mental tests: Review and evaluation. *Monograph of the Society for Research in Child Development,* **30.**

SUCHMAN, J. R. 1960. Inquiry training in the elementary school. *Science Teacher,* **27:** 42–47.

SUTTON-SMITH, B. 1966. Piaget on play: a critique. *Psychological Review,* **73:** 104–10.

THORNDIKE, E. L., and R. S. WOODWORTH. 1901. The influence of improvement in one mental function upon the efficiency of other functions. *Psychological Review,* **8:** 247–61, 384–95, 553–64.

THURSTONE, L. L. 1938. *Primary mental abilities.* Chicago: University of Chicago Press.

TOLMAN, E. C. 1948. Cognitive maps in rats and men. *Psychological Review,* **55**: 189–208.

———, and E. BRUNSWICK. 1935. The organism and the causal texture of the environment. *Psychological Review,* **42**: 43–77.

VERNON, P. E. 1950. The structure of human abilities. New York: John Wiley & Sons, Inc.

———. 1964. Ability factors and environmental influences. *American Psychologist,* **20**: 723–33.

VYGOTSKY, L. S. 1962. *Thought and language,* ed. and trans. by E. Hanfmann and G. Vakar. Cambridge, Mass.: The MIT Press.

WALLACH, M. A. 1963. Research on children's thinking. In National Society for the Study of Education 62nd Yearbook, *Child Psychology.* Chicago: University of Chicago Press.

WARNER, W. L., et al. 1949. *Democracy in Jonesville: A study in quality and inequality.* New York: Harper & Row. (Harper Torchbooks, 1964.)

WARNER, W. L., R. J. HAVIGHURST, and M. B. LOEB. 1944. *Who shall be educated? The challenge of unequal opportunities.* New York: Harper.

WARNER, W. L., M. MEEKER, and K. EELS. 1949. *Social class in America.* Chicago: Science Research Associates.

WEBER, M. 1963. *The sociology of religion.* Boston: Beacon Press.

WELLMAN, B. L., H. M. SKEELS, and M. SKODAK. 1940. Review of McNemar's critical examination of Iowa studies. *Psychological Bulletin,* **37**: 93–111.

WERNER, H., and B. KAPLAN. 1963. *Symbol formation.* New York: John Wiley & Sons, Inc.

WHITE, R. W. 1959. Motivation reconsidered: The concept of competence. *Psychological Review,* **66**: 297–323.

———. 1966. *Lives in progress,* 2nd ed. New York: Holt, Rinehart & Winston, Inc.

WIENER, N. 1948. *Cybernetics.* New York: John Wiley & Sons, Inc.

WOHLWILL, J. F. 1966. Piaget's theory of the development of intelligence in the concrete operations period. *American Journal of Mental Deficiency,* **70** (4): 57–58. Monogr. Suppl.

———. 1967. *Imagination wed to reason.* (A review of Jean Piaget and Bärbel Inhelder. *L'Image mentale chez l'enfant.* Paris: Presses Universitaires de France, 1966.) *Contemporary Psychology,* **12**.

WOLFE, D. 1960. Diversity of talent. *American Psychologist,* **15**: 535–44.

WOLMAN, B. B., ed. 1968. *Historical roots of contemporary psychology.* New York: Harper & Row, Publishers.

WOODWORTH, R. S. 1938. *Experimental psychology.* New York: Holt.

YERKES, R. M. and J. D. DODSON. 1908. The relation of strength of stimulus to rapidity of habit formation. *Journal of Comparative and Neurological Psychology,* **18**: 459–82.

YOUNG, P. T. 1949. Food-seeking drive, affective process and learning. *Psychological Review,* **56**: 98–121.

ZAJONE, R. B. and L. MORRISSETT. 1960. The role of uncertainty in cognitive change. *Journal of Abnormal and Social Psychology,* **61**: 168–75.

ZANKOV, L. V. 1963. Combination of the verbal and the visual in teaching. Trans. by B. Simon, and J. Simon, eds., in *Educational psychology in the U.S.S.R.* Stanford: Stanford University Press.

Index

Similar in form is the *Hermaphroditus and Salmacis* by Thomas Peend published in 1565. Hermaphroditus is the type of Youth tempted by the pleasures of the world who receives the wages of such a sin. Peend had apparently projected a complete translation of the *Metamorphoses* but had been anticipated by Arthur Golding's version of 1565/7. Other Ovidian poems were written in the 1560's, Thomas Underdowne's *Theseus and Ariadne* (1566), Thomas Howell's *Cephalus and Procris* (c. 1568) and William Hubbard's *Ceyx and Alcione* (1569), but Golding is more adequate as an interpreter of Ovid than his contemporaries because his verse does greater justice to Ovid's poetic effects.

In his introduction Golding sufficiently marks out the area that lies between the 'Gothic present' of Ovid moralized and the new poetry. He does regard the *Metamorphoses* as a storehouse of notable moral examples but he does not contemplate producing an Ovid moralized after the French fashion. His reasons are that it would require a book of many quires and that such a labour would be tedious for the author and his readers. A critic, looking before and after Golding, might reasonably conclude that here is clear evidence of a shift in sensibility. If literature for Golding is still a moral activity it need no longer be a moralized one. The poet is assumed to be a moral philosopher without having to expound his doctrine at tedious length. The way is open for Sidney's defence that the poet is the actual creator of the golden or ideal world and for Spenser's attempt to give England a poem at once heroic and philosophical. Golding's merits as poet and translator are considerable and unquestionably helped to make Ovid's poetry once again the gateway to a new age.

Sensibility, however, changes slowly. *The most famous and tragicall historie of Pelops and Hippodamia* (1587) by Matthew Grove can hardly be described as a great advance from the lamentable history of Hubbard. In 1589, however, Thomas Lodge published *Scillaes Metamorphosis* and this poem, although it can be called 'not so much the first poem in a new genre as one of the last in an old one',[14] has certain significant features. The story of Glaucus and Scilla in the *Metamorphoses* involves Circe. Glaucus confesses his love for Scilla to Circe and asks for her help; instead of help she offers her own love and, enraged at its rejection, transforms Scilla in revenge. Lodge transposes this whole scene to the banks of the Isis (which has led some commentators to suggest that he wrote the poem while an undergraduate at Oxford) and Glaucus confesses his love to the poet who is himself wandering love-lorn by the river-bank. Circe does not appear. This changes the whole balance of the poem, for the entire action is now seen through the eyes of the poet. From the beginning of the poem he has been conducting a dialogue with Glaucus on the nature of love; the action is to some extent a demonstration for his benefit, and although the poem is not expounded in moralistic terms it does have a lesson which the poet expects some, at least, of his readers to understand. The only person, according to the poem, really competent to give an account of 'the course of all our plainings' would be 'he that hath seen' the history of Venus and Adonis, Cephalus and Procris, the pangs of Lucina and Angelica the fair. Lodge's ideal spectator is one familiar not only with Ovid but also with Italian epic. The assumption is that the poet is a lover appealing to those who are also in love. The poet comes to understand his own case through his appreciation of the story of Glaucus, and other lovers and their ladies, who are addressed in the Envoy, are offered an increase in self-knowledge. Lodge's mythological parallels and details are there to show this love in all its aspects and 'Furie and Rage, Wan-hope, Dispaire and Woe' exist in this poem for the same reason that they exist in *The Romance of the Rose* and other mediaeval allegorical love-visions. They raise love from a physical to a psychic act. Lodge calls this world of mythology into being because the whole central debate on love can only be understood in its terms. If the actors in the poem, Venus, Glaucus and Scilla herself, are definite in their attitudes and decisive in their actions the poet cannot expect such a final solution in his own case:

> Our talke midway was nought but still of wonder,
> Of change, of chaunce, of sorrow, and her ending;
> I wept for want: he said, time bringes men under,
> And secret want can finde but small befrending.
>   And as he said, in that before I tried it,
>   I blamde my wit forewarnd, yet never spied it.

Lodge's central concern is with the poet rather than the transformation of Scilla and the ironical warning to the ladies. Unlike their moralizing predecessors the Elizabethan Ovidian poets did not claim to know all the answers.

Lodge's new world of mythology was inherited by his successors. A word of caution, however, is necessary. In accounting for a genre of poetry the literary historian is always tempted to treat it as an organic growth with clearly defined origins, flowering and decline. The form is seen as living and dying in the same way as the poets. This is probably